Springer Theses

Recognizing Outstanding Ph.D. Research

Aims and Scope

The series "Springer Theses" brings together a selection of the very best Ph.D. theses from around the world and across the physical sciences. Nominated and endorsed by two recognized specialists, each published volume has been selected for its scientific excellence and the high impact of its contents for the pertinent field of research. For greater accessibility to non-specialists, the published versions include an extended introduction, as well as a foreword by the student's supervisor explaining the special relevance of the work for the field. As a whole, the series will provide a valuable resource both for newcomers to the research fields described, and for other scientists seeking detailed background information on special questions. Finally, it provides an accredited documentation of the valuable contributions made by today's younger generation of scientists.

Theses may be nominated for publication in this series by heads of department at internationally leading universities or institutes and should fulfill all of the following criteria

- They must be written in good English.
- The topic should fall within the confines of Chemistry, Physics, Earth Sciences, Engineering and related interdisciplinary fields such as Materials, Nanoscience, Chemical Engineering, Complex Systems and Biophysics.
- The work reported in the thesis must represent a significant scientific advance.
- If the thesis includes previously published material, permission to reproduce this must be gained from the respective copyright holder (a maximum 30% of the thesis should be a verbatim reproduction from the author's previous publications).
- They must have been examined and passed during the 12 months prior to nomination.
- Each thesis should include a foreword by the supervisor outlining the significance of its content.
- The theses should have a clearly defined structure including an introduction accessible to new PhD students and scientists not expert in the relevant field.

Indexed by zbMATH.

Shun Maegochi

Nonequilibrium Phase Transitions in Driven Vortex Matter

The Reversible-Irreversible Transition, Dynamical Ordering, and Kibble-Zurek Mechanism

Doctoral Thesis accepted by
Tokyo Institute of Technology, Tokyo, Japan

Springer

Author
Dr. Shun Maegochi
Department of Physics
Tokyo Institute of Technology
Tokyo, Japan

Supervisor
Prof. Satoshi Okuma
Department of Physics
Tokyo Institute of Technology
Tokyo, Japan

ISSN 2190-5053 ISSN 2190-5061 (electronic)
Springer Theses
ISBN 978-981-97-2986-9 ISBN 978-981-97-2987-6 (eBook)
https://doi.org/10.1007/978-981-97-2987-6

This Springer imprint is published by the registered company Springer Nature Singapore Pte Ltd.
The registered company address is: 152 Beach Road, #21-01/04 Gateway East, Singapore 189721, Singapore

Supervisor's Foreword

In recent years, there has been a growing interest in nonequilibrium phase transitions that are outside the framework of equilibrium physics. Among them, a variety of ordering phenomena away from equilibrium, which is called dynamical ordering, have been recognized in condensed matter systems. A general interesting question is how the disordered system recovers its order dynamically. Vortices in type-II superconductors are a suitable system to study such phenomena experimentally, since they are regarded as two-dimensional particles interacting with a random substrate, whose dynamics can be studied in a controllable manner.

This book, which is based on the dissertation of Dr. Shun Maegochi, presents nonequilibrium phase transitions associated with dynamical ordering of vortices induced by ac and dc driving forces.

First, a reversible-irreversible transition (RIT) was studied using vortices periodically sheared over a random landscape. Such a vortex system exhibits a transition from a time periodic or reversible phase to a diffusive or irreversible phase with an increase in shear amplitude and vortex density. Critical behaviors around RIT were found to be identical, independent of the driving parameters, indicative of the universality of RIT. In the reversible phase two distinct flow regimes were identified, for the first time, which correspond to a point-reversible state similar to that observed in dilute colloids and a loop-reversible state similar to that found for amorphous systems.

Next, a dynamical ordering transition induced by a dc force was studied. The results demonstrate a second-order dynamical ordering transition from a disordered plastic flow to an anisotropically ordered smectic flow with transverse periodicity. Despite its ubiquitousness, clear experimental evidence of the smectic flow and its evolution as a function of the driving force have not been obtained because of lack of suitable experimental methods. The success of this work was attained by developing new experimental methods to detect transverse order, such as a transverse mode-locking resonance. They will be used in various driven systems, including skyrmions and colloids.

Finally, the author, Maegochi, tested the applicability of the Kibble-Zurek mechanism (KZM) to the dynamical ordering transition from plastic flow to smectic flow. The KZM describes the formation of topological defects in systems crossing a continuous symmetry-breaking phase transition at a finite quench rate. While this mechanism has been extensively studied for equilibrium transitions since 1970s, its applicability to nonequilibrium transitions has not yet been fully examined. Quite recent simulation predicted that the KZM could be applied to dynamical ordering transitions. Maegochi examined the configurational order of vortices in the course of dynamical ordering with various quench rates and found a power-law scaling of defect density with the quench rate. The result, together with the finding of an impulse-adiabatic crossover reflecting critical slowing down, offers the first experimental evidence of KZM in the nonequilibrium transition.

The results obtained in this work indicate that both the RIT and dynamical ordering transition belong to the directed percolation universality class, which is the fundamental class of nonequilibrium phase transitions. Thus, the present findings will be generalized to the broader class of systems that exhibit the nonequilibrium transitions and stimulate further research in them.

Tokyo, Japan
August 2023

Prof. Satoshi Okuma

Parts of this thesis have been published in the following journal articles:

1. S. Maegochi, K. Ienaga, and S. Okuma
   "Evidence of second-order transition and critical scaling for the dynamical ordering transition in current-driven vortices"
   Sci. Rep. **14**, 1232 (2024).
2. S. Maegochi, K. Ienaga, and S. Okuma
   "Kibble-Zurek Mechanism for Dynamical Ordering in a Driven Vortex System"
   Phys. Rev. Lett. **129**, 227001 (2022).
3. S. Maegochi, K. Ienaga, and S. Okuma
   "Moving smectic phase and transverse mode locking in driven vortex matter"
   Phys. Rev. Research **4**, 033085 (2022).
4. S. Maegochi, K. Ienaga, and S. Okuma
   "Critical behavior of density-driven and shear-driven reversible-irreversible transitions in cyclically sheared vortices"
   Sci. Rep. **11**, 19280 (2021).

Acknowledgements

First and foremost, I would like to express my greatest gratitude to my academic advisor, Prof. Satoshi Okuma. This research would not have been possible without his support. From the time I joined the Okuma Lab as a fourth-year undergraduate student, I knew early on that I wanted to enter the doctoral course, and I tried to be proactive in asking questions and engaging in discussions. Professor Okuma was always available for discussions if time allowed, even when he was busy. He treated me as an equal as a researcher, which was a great encouragement. I also learned a great deal from him, especially when it comes to clear and concise writing.

I would also like to thank Dr. Koichiro Ienaga for teaching me many things I needed to know as a researcher, including experimental methods and presentation skills. I am also grateful to Dr. Shin-ichi Kaneko for helpful discussions. I appreciate valuable discussions with all members of Okuma Lab.

I also appreciate helpful examiners' comments from Prof. Toshimasa Fujisawa, Prof. Toru Hirahara, Prof. Fumitaka Kagawa, and Prof. Shuichi Murakami.

Finally, I would like to express my sincere gratitude to my parents and siblings, who have always supported me. My father, who passed away on 10 January 2021, turned my interest to physics and taught me how interesting it is.

This work was supported by a Grant-in-Aid for JSPS Fellows (KAKENHI Grant No. 20J21425) from the Japan Society for the Promotion of Science throughout my Ph.D. course.

Contents

1 Introduction 1
1.1 Nonequilibrium Phase Transition 1
1.1.1 Reversible-Irreversible Transition 1
1.1.2 Absorbing Phase Transition 7
1.1.3 Dynamical Ordering in DC Driven Many-Particle Systems 10
1.2 Metastability and Kinetics 12
1.2.1 Potential Energy Landscape 12
1.2.2 Memory Effect 13
1.2.3 Kibble-Zurek Mechanism 14
1.3 Superconducting Vortex System 16
1.3.1 Classification of Superconductors 16
1.3.2 Vortices in Type-II Superconductors 18
1.3.3 Vortex Flow 20
1.3.4 Peak Effect 21
1.3.5 RIT in the Vortex System 22
1.4 Purposes and Outline of This Thesis 26
References 27

2 Experimental 39
2.1 Sample Preparation 39
2.2 Sample Characteristics 39
2.3 ^{4}He Cryostat 41
2.4 Measurement System 44
2.4.1 Measurements of the Steady-State Voltage 44
2.4.2 Measurements of the Time-Dependent Voltage 46

3 Critical Behavior of RIT Driven by Particle Density as Well as Shear Amplitude 49
3.1 Introduction 49
3.2 Experiments 50

3.3 Results and Discussion 51
3.3.1 The Density-Driven RIT 51
3.3.2 Two Types of Reversible States 55
3.4 Conclusion 57
References 58

4 Moving Smectic Phase and Transverse Mode Locking in Driven Vortex Matter 61
4.1 Introduction 61
4.2 Experiments 62
4.2.1 Two-Step Measurements 62
4.2.2 Mode-Locking Measurements 63
4.3 Results and Discussion 64
4.3.1 Two-Step Measurements 64
4.3.2 Transverse and Longitudinal Mode Locking 68
4.4 Conclusion 70
References 70

5 Evidence of Second-Order Transition and Critical Scaling for the Dynamical Ordering Transition 73
5.1 Introduction 73
5.2 Experiments 73
5.3 Results and Discussion 74
5.3.1 Transverse Depinning 74
5.3.2 Dynamical Ordering Transition 75
5.3.3 Scaling Analysis 77
5.4 Conclusion 80
References 80

6 Kibble-Zurek Mechanism for the Dynamical Ordering Transition 83
6.1 Introduction 83
6.2 Experiments 84
6.3 Results and Discussion 86
6.4 Conclusion 89
References 91

7 Conclusions 95
References 98

Curriculum Vitae 99

Chapter 1
Introduction

Systems out of equilibrium are abundant in nature as notably represented by life and they are of great interest in many aspects. One of the important aspects of nonequilibrium systems is that the dynamics of the system gives rise to various ordered phases via self-organizing processes [1–7]. In nature, dynamics often emerges intrinsically as in active self-propelled materials and biological systems [8–16], which exhibit peculiar phenomena such as flocking [17–21] and motility-induced phase separation [22–27]. Of course, the dynamics is also induced extrinsically by applied driving forces. In this thesis, we focus on the latter case, that is, dynamics of externally driven many-particle systems.

1.1 Nonequilibrium Phase Transition

In the last decades, the dynamics of many-particle systems subjected to external driving forces has attracted much research interest because they exhibit a variety of nonequilibrium phases and phase transitions, such as a depinning transition [28–49] and a reversible-irreversible transition (RIT) [50–79]. In the following sections, we outline the background of dynamical ordering phenomena and related nonequilibrium phase transitions in many-particle systems driven by either ac or dc force.

1.1.1 Reversible-Irreversible Transition

When assemblies of many particles with a random initial configuration are subjected to periodic shear, the particles collide to each other and change their arrangement from one cycle to the next. With increasing the number of shearing cycles, particles spontaneously evolve into an organized configuration where less collision occurs. This self-organizing process is named "random organization" [53, 59, 80–88], which

S. Maegochi, *Nonequilibrium Phase Transitions in Driven Vortex Matter*, Springer Theses,
https://doi.org/10.1007/978-981-97-2987-6_1

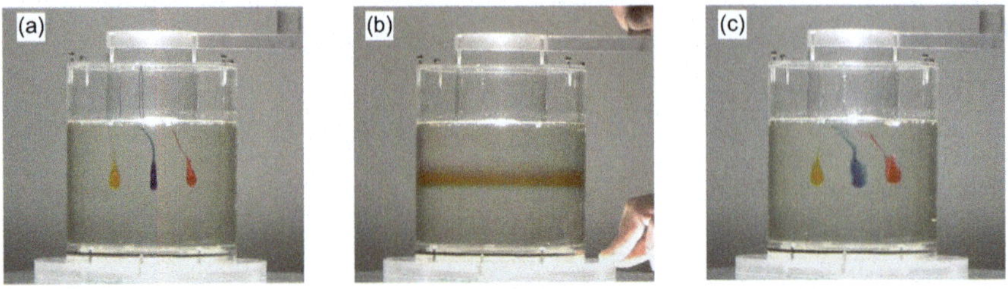

Fig. 1.1 Experimental demonstration of the reversible, laminar flow. **a** The droplets of yellow, blue, and red dye injected into the viscous fluid are placed in the gap between two concentric cylinders. **b** The system is sheared and the dye seems to be intermixed. **c** The system is sheared back and the dye returns to its starting positions. The original movie is available on YouTube: (https://www.youtube.com/watch?v=_dbnH-BBSNo)

has been observed experimentally as well as numerically in various many-particle systems, such as colloidal suspensions [53], soft glasses [63], dislocations [82], skyrmions [74], and vortices in type-II superconductors [54, 59, 71, 75, 79]. Associated with the random organization, a novel nonequilibrium phase transition from a reversible flow to an irreversible flow called RIT was reported with an increase in strain amplitude [51–53].

A well-known example of the reversible flow is a basic shear flow (pipe, channel, and Couette flows) of a Newtonian fluid at low Reynolds number (large viscosity). Figure 1.1 shows the apparatus called a Couette cell used for the visual examination of the reversible flow, especially known as a laminar flow [89]. This apparatus consists of two concentric cylinders, which contain droplets of yellow, blue, and red dye as well as viscous fluid in between the cylinders as shown in Fig. 1.1a. They are initially in a state of equilibrium and the diffusion of dye due to Brownian motion is almost negligible because of the large viscosity of the fluid. As shown in Fig. 1.1b, when the inner cylinder is slowly rotated, the fluid is dragged and a global shear which is roughly inversely proportional to the radius is applied to the system. After several rotations of the inner cylinder, the dye seems to be completely intermixed. However, after the same amount of rotations in the opposite direction, as shown in Fig. 1.1c, the dye almost returns to its starting positions. This reversible behavior is consistent with the time-reversal symmetry of the dynamical equations known as the Stokes equations [52]. Similarly, slowly sheared suspensions of non-Brownian particles like colloids are governed by time reversible equations of motion.

In 2005, Pine et al. experimentally studied how time reversibility fails for slowly sheared colloidal suspensions [51]. They used the similar apparatus explained above and found that the periodically sheared suspension has an onset of irreversibility depending on the magnitude of the strain amplitude γ. Figure 1.2a shows displacement vectors of particles along the flow direction (x) and the axial direction (z) after one cycle. The particles do not return to their starting positions and exhibit the irreversible motion. Figure 1.2b shows chaotic particle trajectories sampled over many periods, indicating the diffusive motion. For quantitative analysis, they calculated the

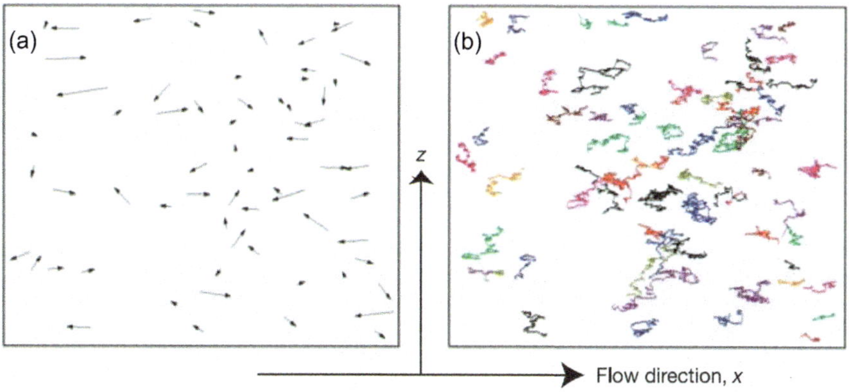

Fig. 1.2 Particle displacements and trajectories above the onset of irreversibility. **a** Particle displacements along the flow direction (x) and the axial direction (z) after one cycle sheared. **b** Irreversible particle trajectories sampled over several periods. Reprinted with permission from Springer Nature, Ref. [51]

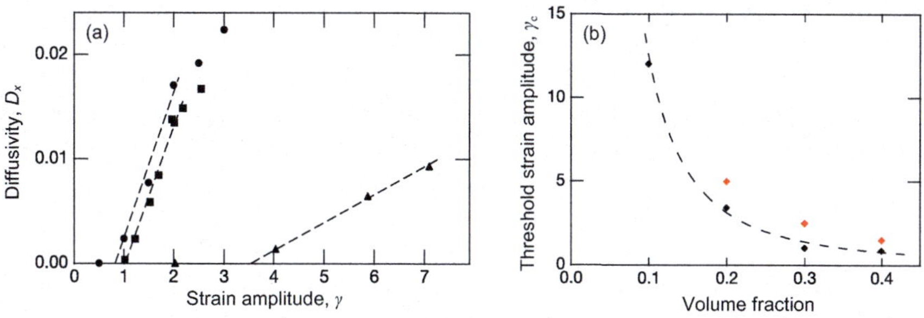

Fig. 1.3 Diffusivity and threshold strain amplitude. **a** Diffusivity D_x versus strain amplitude γ for different particle volume fractions: triangles, $\phi = 0.20$; squares, $\phi = 0.30$; circles, $\phi = 0.40$. Dashed lines indicate the linear extrapolations to zero diffusivity. **b** Threshold strain amplitude versus volume fraction. The experimental threshold amplitudes (black diamonds) determined from linear extrapolation in (**a**) are consistent with the numerical ones (red diamonds). Adapted with permission from Springer Nature, Ref. [51]

diffusivity of the particles in the flow direction, $D_x = \langle \Delta x^2 \rangle / 2$, where $\langle \Delta x^2 \rangle$ is the mean square particle displacement per cycle. In Fig. 1.3a, D_x is plotted as a function of strain amplitude γ for different particle volume fractions. For small γ, the diffusivity is zero and the flow is reversible, while for large γ above a threshold γ_c, the diffusivity and irreversibility emerge due to the persistent particle collisions. Figure 1.3b shows that the threshold γ_c, which corresponds to RIT, decreases as the volume fraction increases because of an increase in the frequency of particle collisions.

In 2008, Corté et al. carried out the further work on this system and found for the first time that the transition from reversible to irreversible states is the nonequilibrium phase transition [53]. Let us review their numerical and experimental studies for the

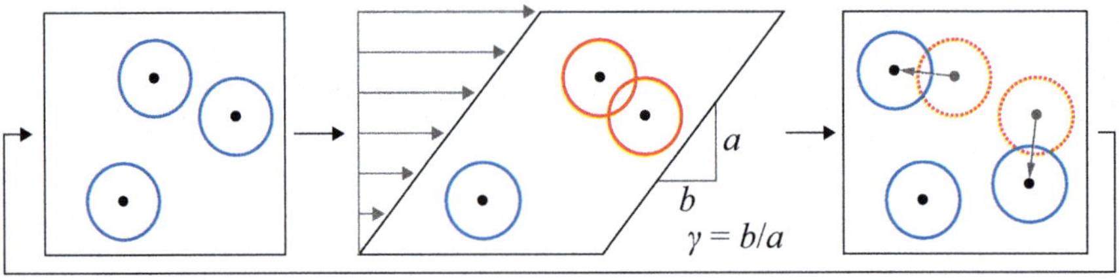

Fig. 1.4 Schematic illustration of one cycle of the collision model [53]. When the shear is applied to the system, the particles that pass close to each other collide and are displaced by a small distance in random direction after one cycle

colloidal system. The idea of their simulation model is simple such that when the shearing force is applied to the suspension, the particles that pass close to each other collide and are displaced by a small distance after one cycle. Figure 1.4 schematically illustrates this process. First, we consider a two-dimensional (2D) rectangular box containing randomly distributed particles. Then the system is sheared back and forth with the strain amplitude γ. The definition of γ is shown in the figure. The particles that pass within a distance d, corresponding to a particle radius, will collide as indicated by overlapping red circles in the figure. After removing the shear, collided particles are displaced in random direction with a small displacement. Particles that do not collide return to their initial positions. The number of colliding particles increases with increasing γ and the particle volume fraction.

Figure 1.5a, b show the snapshots of the particle configuration evolved from the random initial distribution for $\gamma = 3$ and $\gamma = 2$, respectively, as a function of the number of shear cycles. Since the initial particle distribution is random, many particles collide during the first few cycles for both γ, as indicated by a large number of black dots. After several hundred cycles, however, the particles exhibit different behaviors for each γ. For $\gamma = 2$, as shown in Fig. 1.5c, the fraction of active particles per cycle $f_{\mathrm{a}}(t)$ decreases with increasing the number of shear cycles and vanishes after a long time. This behavior is also depicted by no black dots in the last snapshot in Fig. 1.5b, resulting from random organization. Through colliding with each other, all the particles have explored a different configuration in which they no longer collide when the system is sheared. Once the system falls into such a state, any changes in particle configuration never occur and the system is unable to escape from it. In that sense, this state corresponds to a so-called absorbing phase.

On the other hand, for $\gamma = 3$, the system settles into an irreversible steady state in which some fraction of particles always collide with other particles. In this case, $f_{\mathrm{a}}(t)$ decreases to a steady-state value $f_{\mathrm{a}}^{\infty}(\neq 0)$ and fluctuates indefinitely. Hence, this state corresponds to a so-called active phase. The relaxation curves in the main panel of Fig. 1.5c are satisfactory fitted by the following function:

$$f_{\mathrm{a}}(t) = (f_{\mathrm{a}}^{0} - f_{\mathrm{a}}^{\infty})\frac{e^{-t/\tau}}{t^{\alpha}} + f_{\mathrm{a}}^{\infty}. \tag{1.1}$$

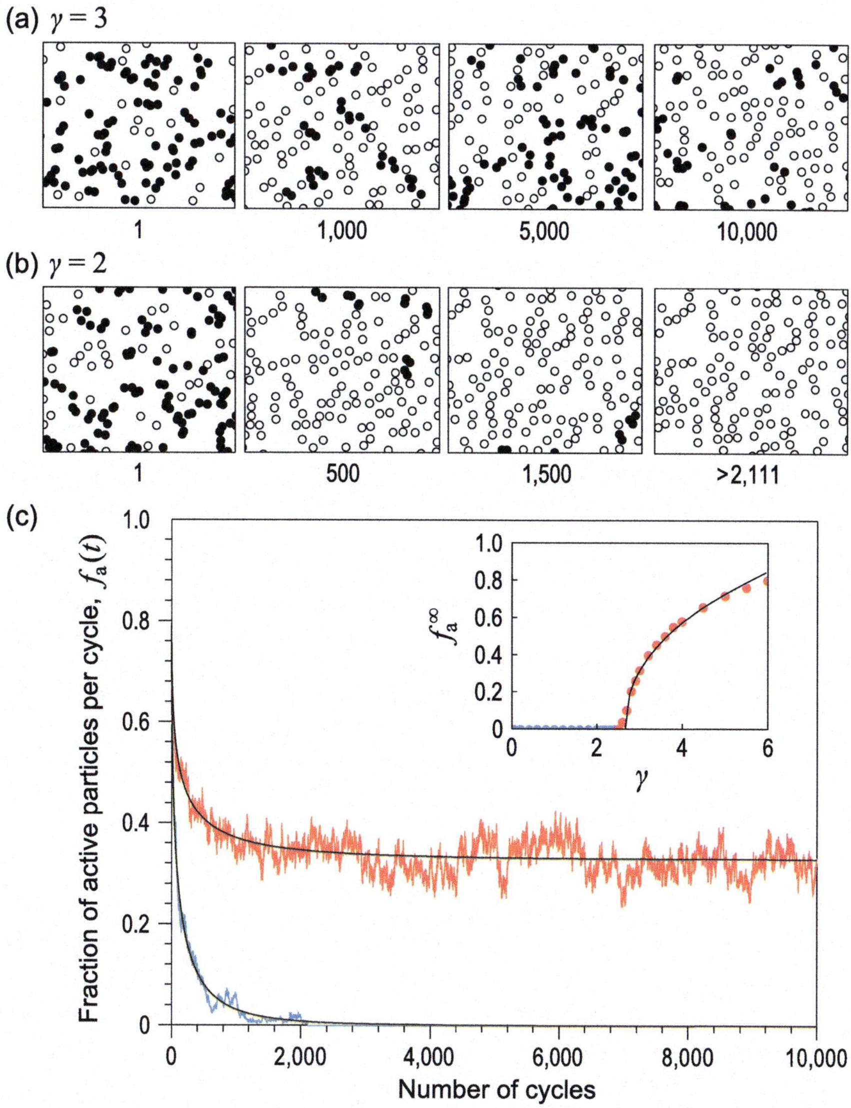

Fig. 1.5 Simulation results for 2D model. **a**, **b** Snapshots of the particle configuration evolved from random initial distributions for strain amplitude $\gamma = 3$ (**a**) and $\gamma = 2$ (**b**). The number of shear cycles is shown below each snapshot. Black dots represent irreversible, active particles that will collide in the next one cycle and open circles represent reversible particles. **c** Fraction of active particles per cycle $f_a(t)$ versus number of shear cycles for $\gamma = 3$ (red) and $\gamma = 2$ (blue). The full lines indicate the fits to Eq. 1.1. Inset: The steady-state value of $f_a(t)$ as a function of γ. Adapted with permission from Springer Nature, Ref. [53]

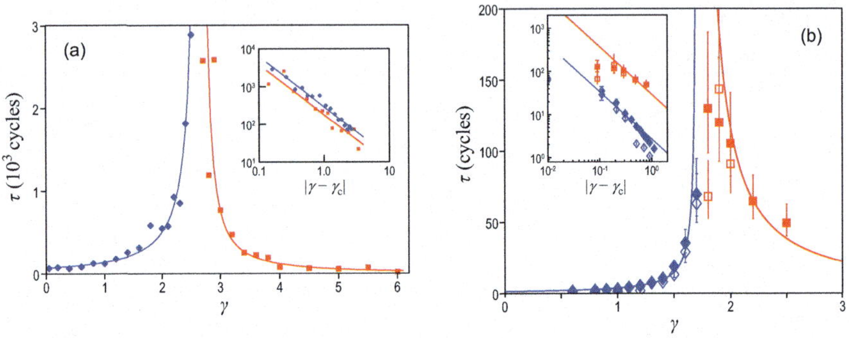

Fig. 1.6 **a** Simulation and **b** experimental results for the relaxation time τ as a function of the strain amplitude γ. Blue and red symbols are the data below ($\gamma < \gamma_c$) and above ($\gamma > \gamma_c$) RIT, respectively. $\gamma_c = 2.66 \pm 0.05$ for the simulation and $\gamma_c = 1.71 \pm 0.15$ for the experiment. Full lines indicate the power-law fits to $\tau \propto |\gamma - \gamma_c|^{-\nu}$ with the critical exponent $\nu = 1.33 \pm 0.02$ (**a**) and $\nu = 1.1 \pm 0.3$ (**b**). Insets of (**a**) and (**b**): Replots of the data shown in the main panel, as τ versus $|\gamma - \gamma_c|$, on a double-logarithmic scale. Adapted with permission from Springer Nature, Ref. [53]

Here, f_a^0 and f_a^∞ are the initial and steady-state values of $f_a(t)$, respectively, and τ is the characteristic time at which the relaxation crosses over from a power law with an exponent α to an exponential form. The inset of Fig. 1.5c shows how f_a^∞ increases with γ. A continuous transition from the reversible, absorbing phase ($f_a^\infty = 0$) to the irreversible, active phase ($f_a^\infty > 0$) takes place at a threshold amplitude $\gamma_c = 2.66 \pm 0.05$. The fraction of active particles in the steady state f_a^∞ serves as an order parameter and obeys the power law above the transition, $f_a^\infty \propto (\gamma - \gamma_c)^\beta$, with the critical exponent $\beta = 0.45 \pm 0.02$.

The relaxation time τ required for the system to reach the steady state obtained from Eq. 1.1 with $\alpha = 0.15 \pm 0.05$ exhibits a power-law divergence at the same $\gamma_c \simeq 2.66 \pm 0.05$ as shown in Fig. 1.6a. The divergence on both sides of the transition is fitted to $\tau \propto |\gamma - \gamma_c|^{-\nu}$ with the critical exponent $\nu = 1.33 \pm 0.02$. As a result, the simulation model predicted that RIT is the nonequilibrium dynamical phase transition accompanied with transient behavior. To demonstrate this prediction, Corté et al. also conducted the experiment on colloidal suspensions. Figure 1.6b displays the experimental result of the power-law divergence of $\tau(\gamma)$. This divergence on both sides of the transition at $\gamma_c = 1.71 \pm 0.15$ can be fitted to $\tau \propto |\gamma - \gamma_c|^{-\nu}$ with the critical exponent $\nu = 1.1 \pm 0.3$. The experimental result demonstrates the realization of RIT in the actual system. The difference in values of ν is due to the different dimensionality for the 2D simulation and the 3D experiment. The values of ν are, within errors, consistent with theoretical values for a directed percolation universality class explained in the next section.

1.1.2 Absorbing Phase Transition

It was theoretically predicted that a lot of nonequilibrium phase transitions including RIT may be categorized into the universality class of the absorbing phase transition [90–94]. In the absorbing phase, the externally driven system organizes into a state where the fluctuations are strongly suppressed or even completely eliminated. Once the system falls into such a state, the system is unable to leave it and hence detailed balance is broken, i.e., this is a nonequilibrium process [90]. On the other hand, the system in the active phase reaches a steady state where the particle activity survives permanently. Actually, a variety of systems in nature exhibit the transition between active and absorbing phases, so that many theoretical models have been proposed. In particular, the best studied and simplest model for the absorbing phase transition, as prominent as the Ising model for equilibrium phase transitions, is directed percolation (DP) introduced in 1957 by Broadbent and Hammersley [95], which is an anisotropic modification of ordinary percolation. In the following, we give a brief introduction to DP with some examples in nature.

Figure 1.7 illustrates typical representations of ordinary isotropic bond percolation (a) and directed bond percolation (b) [96]. In these models, the sites of two-dimensional diagonal square lattice are connected by bonds (channels). The bonds are open (connected) with probability p and otherwise closed (disconnected), as shown with solid and dotted lines, respectively. Such a situation can be realized in a porous medium where neighboring pores are connected by channels of different permeability. In ordinary percolation [97], when we inject the water into a single site, the water propagates isotropically in all directions through open bonds. If the probability p is sufficiently large, the water will percolate through the medium. On the other hand, if p is small enough, the propagation of the water will stop without

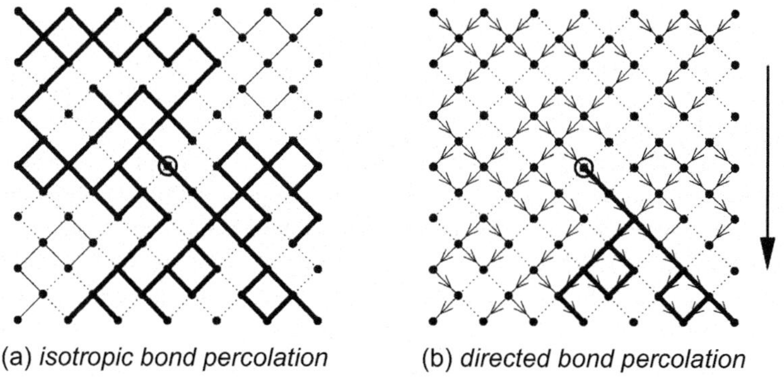

Fig. 1.7 Isotropic and directed bond percolation models on a two-dimensional diagonal square lattice. **a** Ordinary isotropic percolation. **b** Directed percolation. Arrows indicate the direction of propagation. In both cases, open (connected) and closed (disconnected) bonds are represented by solid and dotted lines, respectively. Bold lines indicate the percolating cluster generated from the central site. Reprinted from Ref. [96], Copyright 2001, with permission from Elsevier

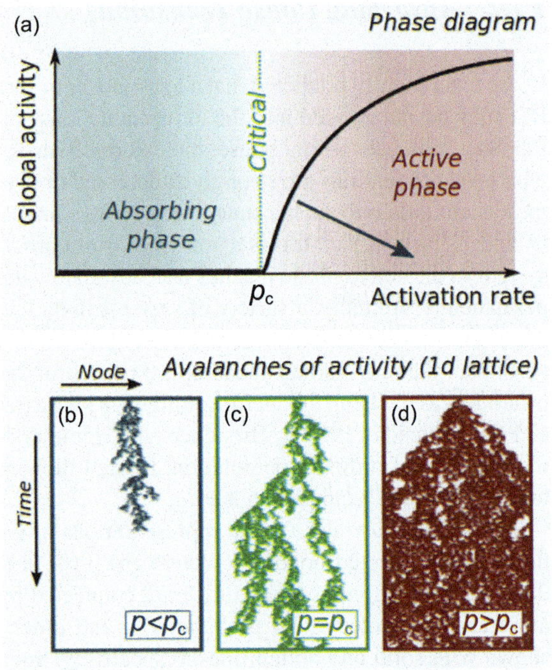

Fig. 1.8 **a** Phase diagram of the absorbing phase transition. **b–d** Space-time histories of spreading process from a single active seed for different spreading rates p. **b** $p < p_c$, the absorbing phase. **c** $p = p_c$, at criticality. **d** $p > p_c$, the active phase. Colored dots indicate active sites (e.g., burning trees for the forest fire). Horizontal axis shows the row of the sites. Vertical axis indicates time, that is, the spreading direction. Adapted figure with permission from Ref. [98]. Copyright 2018 by the American Physical Society

percolation. Both regimes are separated by a continuous phase transition at a certain threshold p_c.

Here, the survival probability P_∞ of whether a randomly selected site belongs to a percolating cluster (bold lines in the figure) serves as an order parameter and follows $P_\infty(p) \propto (p - p_c)^{\beta'}$ for $p > p_c$, while it is zero below the transition $p < p_c$. In the case of DP, a specific direction is introduced to the water propagation, as indicated by arrows in Fig. 1.7b. This model mimics the porous medium in a gravitational field where the water can only flow downwards. Similar to isotropic percolation, DP is characterized by the order parameter P_∞.

Another example of DP behavior is the spreading of the forest fire. In this case, the fire is assumed to spread to neighboring trees with the probability p. Figure 1.8b–d show typical space-time histories of spreading flame from a single active seed at the top [98]. The colored dots indicate the burning trees (active sites) and the other regions are the absorbing or quiescent state. For $p < p_c$ as shown in Fig. 1.8b, the flame is hard to spread, so the number of burning trees decreases and eventually vanishes. This state is in the absorbing phase. In contrast, Fig. 1.8d shows that the system for $p > p_c$ enters the active phase, in which the forest is stationary burning. Figure 1.8c shows the configuration at criticality $p = p_c$, where the average number of active sites $N_a(t)$ increases asymptotically as $N_a(t) \propto t^\theta$. It is noteworthy that when the spreading process starts from fully active sites instead of the single seed, the fraction of active sites $\rho_a(t)$ at criticality decays asymptotically as

$$\rho_a(t) \propto t^{-\alpha} \quad (1.2)$$

with the critical exponent α. Here again we can define the order parameter ρ_a^∞ as the steady-state value of $\rho_a(t)$ and above the transition it obeys the power law

$$\rho_a^\infty \propto (p - p_c)^\beta, \quad (1.3)$$

where β is the critical exponent. Figure 1.8a illustrates the schematic phase diagram.

In addition, the space-time histories of spreading process are characterized by a spatial correlation length $\xi_\perp$ and a temporal correlation length $\xi_\parallel$. The meaning of these length scales are as follows. In the absorbing phase shown in Fig. 1.8b, the lateral size and the lifetime of clusters originating from the single active seed are proportional to $\xi_\perp$ and $\xi_\parallel$, respectively. In the active phase shown in Fig. 1.8d, the opening angle of the cluster is determined by the ratio $\xi_\perp/\xi_\parallel$. We note that the temporal correlation length $\xi_\parallel$ corresponds to the relaxation time τ explained in the previous section. Close to the transition, these length scales show the power-law divergence as

$$\xi_\perp \propto |p - p_c|^{-\nu_\perp}, \quad \xi_\parallel \propto |p - p_c|^{-\nu_\parallel} \quad (1.4)$$

with critical exponents $\nu_\perp$ and $\nu_\parallel$, respectively. And also, two length scales are related by $\xi_\parallel \propto \xi_\perp^z$, where $z = \nu_\parallel/\nu_\perp$ is known as the dynamical exponent. The estimates for critical exponents of DP are listed in Table 1.1.

Besides above examples, a lot of models such as epidemics [99], turbulence [100], catalytic reactions [101–103], formation of galaxies [104, 105], flowing sand [106], vortices in type-II superconductor [36] exhibit the absorbing phase transition in the DP universality class. Despite these considerable theoretical efforts, experimental realization of DP critical behavior, especially precise determination of the critical exponents are poor. This is pointed out by Grassberger [107]: "... there is still no experiment where the critical behavior of DP was seen. This is a very strange situation in view of the vast and successive theoretical efforts made to understand it. Designing and performing such an experiment has thus top priority in my list of open problems."

Table 1.1 Critical exponents of directed percolation obtained by numerical methods [90]

Critical exponent	$D = 1$	$D = 2$	$D = 3$
$\beta = \beta'$	0.276486(8)	0.584(4)	0.81(1)
$\nu_\perp$	1.096854(4)	0.734(4)	0.581(5)
$\nu_\parallel$	1.733847(6)	1.295(6)	1.105(5)
z	1.580745(10)	1.76(3)	1.90(1)
$\delta = \alpha$	0.159464(6)	0.451	0.73
θ	0.313686(8)	0.230	0.12

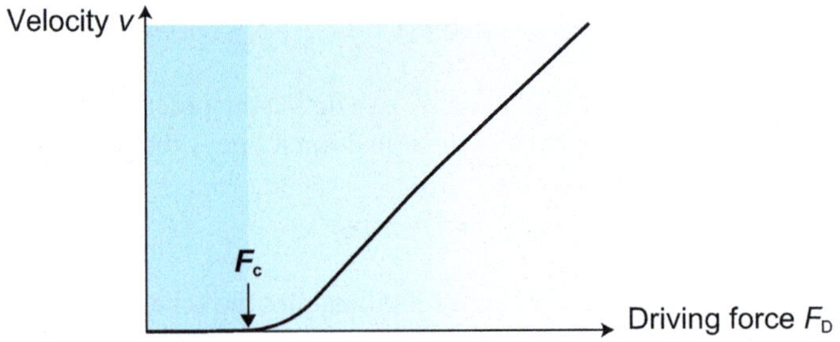

Fig. 1.9 Schematic illustration of the driving force-velocity curve. F_c marks the depinning

In 2007, Takeuchi et al. experimentally observed the critical behavior of DP in nematic liquid crystals [108]. They investigated the nonequilibrium phase transition between two topologically different turbulent states and obtained as many as 12 critical exponents. Furthermore, they confirmed 5 scaling functions and 8 scaling relations, which are in full agreement with those theoretically expected for the DP universality class [109, 110]. Thus, they first revealed the appearance of DP criticality in nature.

Recently, several experimental observations of DP criticality have been reported in turbulent liquid crystals [108–110], driven suspensions [53], basic shear flows [111, 112], cold atoms [113], volatile drops [114], and vortices in type-II superconductors [45, 46, 49, 59, 71, 75]. As mentioned before, RIT is considered as a new family of DP.

1.1.3 Dynamical Ordering in DC Driven Many-Particle Systems

In Sect. 1.1.1, we have reviewed the organizing process in periodically driven systems. Here we focus on ordering phenomena induced by the dc drive instead of the ac drive [28]. Let us consider assemblies of many particles driven over a random substrate, where particles interact repulsively with each other. The substrate has many potential energy minima called pinning centers, which exert the attractive force to the particles around them. The repulsive interaction between particles promotes the formation of the ordered configuration, e.g., a triangular lattice, while the particle-pinning interaction hinders this ordering. The balance between these competing effects is altered by applying the driving force. Figure 1.9 shows a schematic of the driving force-velocity ($F_D - v$) relation. When the driving force is small, the particles are pinned and their velocity is zero. With an increase in F_D, particles undergo a depinning transition [28–49] at a depinning threshold F_c and initiate the flow. Just above the depinning, particles are set in the disordered flow dominated by pinning termed a plastic flow, where v increases nonlinearly. With further increasing

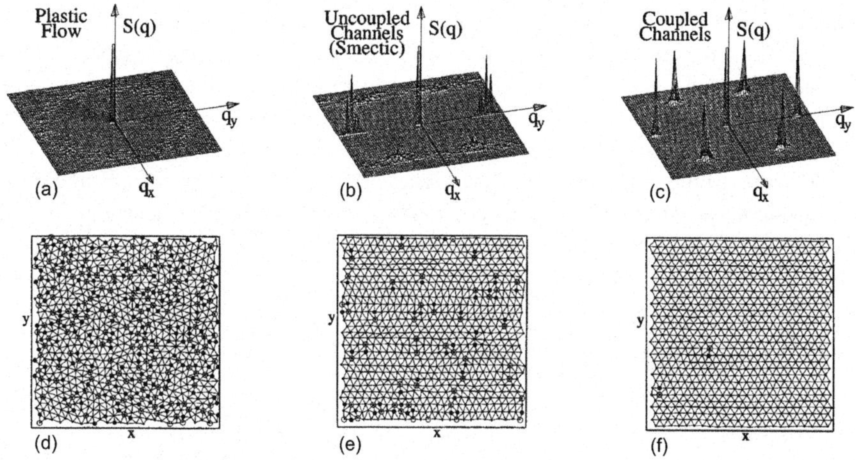

Fig. 1.10 **a–c** Structure factor S($\boldsymbol{q}$) and **d–f** Delaunay triangulations from the 2D vortex simulation. **a, d** The plastic flow. **b, e** The moving smectic phase. **c, f** The moving Bragg glass. In Delaunay triangulation images, vortices are placed at vertices of triangles. Solid (open) circles indicate fivefold (sevenfold) coordinated vortices, i.e., topological defects in the triangular lattice. The vortices are driven in the x direction. Adapted figure with permission from Ref. [125]. Copyright 1998 by the American Physical Society

F_D, the coupling between the particles and the pinning centers is reduced and the pinning becomes ineffective. Consequently, the disordered flow finally turns into the ordered flow dominated by the particle-particle interaction. This ordering phenomenon is known as dynamical ordering [28]. The crossover from the nonlinear to linear increase in $v(F_D)$ reflects the dynamical ordering.

A wide variety of systems exhibit the dynamical ordering, such as moving Wigner crystals [115], charge-density waves [116], colloidal particles [117], and magnetic skyrmions [118]. Among them, vortices in type-II superconductors are the most well studied system both theoretically and experimentally over three decades [119–139]. As shown in Fig. 1.10, a molecular dynamics simulation for the 2D vortex system explored the structure of vortices through the dynamical ordering [125]. Figure 1.10a–c are the structure factor $S(\boldsymbol{q})$, where Bragg peaks indicate the presence of orientational order, while Fig. 1.10d–f show corresponding Delaunay triangulations, where vortices are placed at vertices of triangles and circles indicate topological defects (dislocations) in the triangular lattice. The vortices are driven in the x direction.

At low driving force, in the plastic flow regime shown in Fig. 1.10a, d, no peaks in $S(\boldsymbol{q})$ and the large number of defects reveal the disordered flow of vortices. At high driving force regimes, there are two different types of ordered states, termed a moving smectic phase (uncoupled channels) shown in Fig. 1.10b, e and a moving Bragg glass (coupled channels) shown in Fig. 1.10c, f. In the moving smectic phase, $S(\boldsymbol{q})$ has well-defined peaks only along q_y axis, namely, along the direction transverse to the drive. The pair of defects in the vortex lattice are as well oriented transverse to the drive. Vortices are flowing in channels, which are parallel to the driving direction x and almost regularly spaced transverse to the driving direction. Adjacent channels that are uncoupled easily slip past each other. Thus, the moving smectic phase is characterized by anisotropic structures with long-range order transverse to the driving direction. At much higher driving force, the moving Bragg glass regime appears, where all channels are coupled and vortices reorder into the nearly defect-free isotropic triangular lattice with sixfold Bragg peaks.

1.2 Metastability and Kinetics

In the previous section, we have seen the spontaneous emergence of order by dynamics of interacting particles in nonequilibrium systems. Another intriguing aspect of nonequilibrium process is the presence of flux of energy and the absence of energy conservation, which enables the system to reach metastable hidden ordered states that are typically not accessible under thermal equilibrium conditions [140–157]. This accessibility is believed to open a new avenue for fabrication of devices with unimaginable functionalities [158]. For understanding metastability and related phenomena, such as memory formation in matter [159], it is useful to consider a rugged free-energy or potential-energy landscape [160–171]. Here, we provide the basic idea of the energy landscape and its application to the nonequilibrium, kinetic process.

1.2.1 Potential Energy Landscape

For a 3D system consisting of N particles, one can define the potential energy for each configuration of particles $\boldsymbol{R} = (\boldsymbol{r}_1, \boldsymbol{r}_2, \ldots, \boldsymbol{r}_N)$, which is mapped onto each point of a $3N$-dimensional hypersurface $U(\boldsymbol{R})$. This multidimensional potential energy surface is called the potential energy landscape (PEL) [161, 162]. Figure 1.11a shows a schematic 1D illustration of PEL, where the horizontal axis indicates all configurational coordinates in order to briefly visualize the multidimensional PEL. The kinetics of the system is viewed as the motion of the point on PEL as shown in Fig. 1.11b. For example, the left minimum is a stable state, i.e., the perfect crystal, while the right local minimum or a basin is a metastable state. The particle configuration and potential energy between these minima are similar, however, particles shown with blue and yellow are swapped. For exchanging neighbors, particles should break bonds and rearrange as shown in the lower middle panel, and then the system have to surmount the potential energy barrier (saddle point) between the minima. Note that the system at each minima or basins is in a state of thermal equilibrium and nonequilibrium nature appears within the activation process.

The idea of PEL has been widely employed in fields from physics and chemistry to the information science and robotics. Examples include the liquid-glass

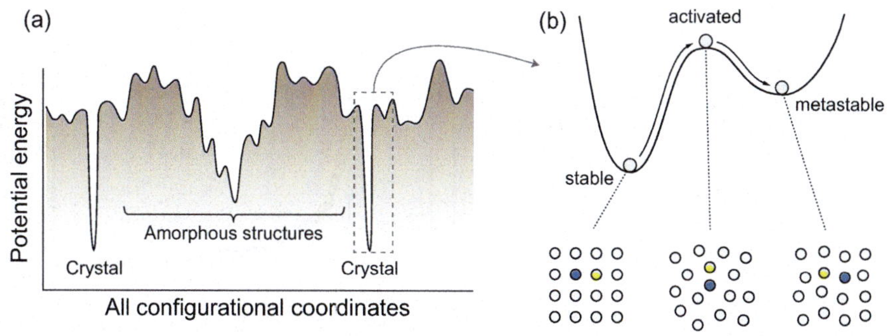

Fig. 1.11 **a** Schematic 1D illustration of PEL for a many-particle system. The horizontal axis indicates all configurational coordinates. **b** Enlarged view of the dashed region in (**a**). The left and right minima respectively represent the stable and metastable states, while the saddle point represents the unstable, activated state. Corresponding particle configurations are shown in the lower panel

transition [172–175], complex behavior of (spin-)glasses [176–179], deformation of solids [180–182], chemical reactions [183, 184], protein folding [185–191], terrestrial locomotion [192, 193], and machine learning [194, 195].

1.2.2 Memory Effect

There are many forms of memory around us, such as memory storage in computers, immune memory by vaccine [196–199], and memory in our brain: for example, the readers of this thesis have short-term memory of what was discussed in the previous section, that is, the potential energy landscape. The memory will be lost when the system is completely relaxed to the thermal equilibrium steady state, corresponding to the global minimum of PEL. Therefore, many types of memory are connected with nonequilibrium behavior and transient response to stimuli [200–212]. Although there are a myriad of examples of memory and history effect, we leave details to the review article [159] and briefly introduce flow induced memory of vortex systems.

Figure 1.12 shows the time evolution of voltage (average vortex velocity) shown with open squares in response to current (driving force) shown with lines [137]. When the current pulse is first applied to the ordered vortex system, the voltage decreases, indicating that moving vortices gradually become a less ordered configuration. As the current is turned off, the voltage abruptly drops to zero, namely, vortex motion is frozen. When the current is restored after a waiting time, the decay of the voltage restarts with exactly the same time dependence from the point where it stopped. The result indicates that the system stores the memory of its flow state prior to being stopped. It is attributed to the existence of the metastable state in which the system is trapped during the waiting time in the absence of current. Thus, the memory effect

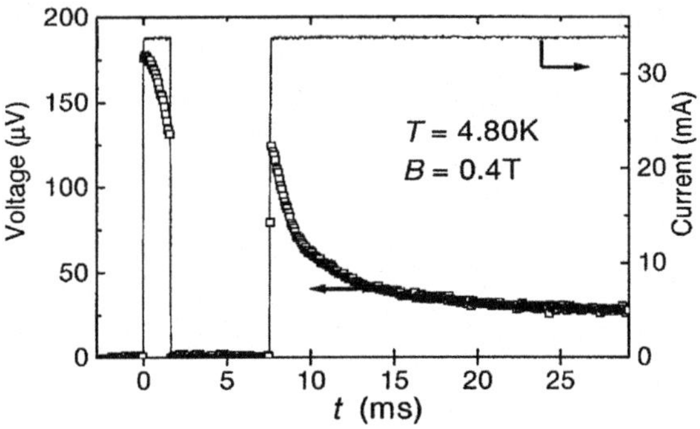

Fig. 1.12 Memory of the vortex flow state. Open squares and lines are the time evolution of the voltage (average vortex velocity) and the current (driving force), respectively. Reprinted figure with permission from Ref. [137]. Copyright 1999 by the American Physical Society

enables us to retrieve information on the past flow state of vortices by applying the current that serves as a probe [71, 213–215].

1.2.3 Kibble-Zurek Mechanism

In this section, we consider the metastability related to symmetry-breaking phase transitions that are ubiquitous in nature. For instance, when water is cooled below freezing point, the rotational symmetry is broken and an ice crystal with a certain orientation is obtained. When the cooling is slow enough, the resulting crystal contains no defects because the system samples PEL entirely and finds the most stable ordered structure. On the other hand, the system subjected to the fast quench often falls into the local minimum of PEL, namely, the metastable state, where the crystal structure is disordered with many defects or completely amorphous. The density of defects would increase with increasing the quench rate. In particular, for continuous second-order phase transitions, the relation between the defect density and the quench rate is formulated as the Kibble-Zurek mechanism (KZM) [216–218], which was originally proposed by Kibble for the cosmological phase transition in the early Universe [219, 220] and refined by Zurek [221, 222] for the condensed matter systems.

In general, as shown in Fig. 1.13, the second-order phase transition is characterized by the divergence of equilibrium spatial correlation length ξ and equilibrium relaxation time τ as a function of reduced control parameter (dimensionless distance from the critical point) ϵ,

$$\xi(\epsilon) \propto |\epsilon|^{-\nu}, \quad \tau(\epsilon) \propto |\epsilon|^{-\nu z}, \tag{1.5}$$

where ν and z are correlation-length and dynamical critical exponents, respectively [223] (ν and νz correspond to $\nu_\perp$ and $\nu_\parallel$ in Sect. 1.1.2, respectively). For thermal phase transitions, $\epsilon = (T_c - T)/T_c$, where T_c is a critical temperature. When the system crosses the transition ($\epsilon = 0$), time dependence of ϵ in the vicinity of the critical point can be linearized as $\epsilon(t) = t/t_q = \dot{\epsilon}t$, where t_q is the quench time that determines the change rate of $\epsilon(t)$ and $\dot{\epsilon}$ stands for time derivative of $\epsilon(t)$, that is, the quench rate.

According to the KZM, the system goes through three regimes, termed adiabatic, impulse or frozen-out, and adiabatic, during the quench as shown in Fig. 1.13a. Far away from the critical point, $\epsilon(t) < 0$ (or $t < 0$), the relaxation time $\tau(\epsilon)$ shown with solid lines is smaller than the remaining time $|t|$ to reach the critical point shown with dashed lines, so that the system can follow change of $\epsilon(t)$ and hence the dynamics is adiabatic. This stage is usually referred to as the adiabatic regime. In the vicinity of the transition at $\epsilon(t) \approx 0$, $\tau(\epsilon)$ becomes too large for the system to relax into the state of current $\epsilon(t)$ because of the critical slowing down and the dynamics is approximately frozen. This stage is called the impulse regime or the frozen-out regime. For $\epsilon(t) > 0$, the dynamics resumes and becomes adiabatic again when $\tau(\epsilon)$ becomes small with respect to the time t elapsed after crossing the transition. This simplified scenario of the KZM is known as an adiabatic-impulse-adiabatic approximation [218].

Boundaries between adiabatic-impulse regimes and between impulse-adiabatic regimes are estimated by comparing $\tau(\epsilon)$ and an inverse transition rate $|\epsilon(t)/\dot{\epsilon}(t)|(= |t|)$. Intersection points of solid and dashed lines in Fig. 1.13a are given by the equation

$$\tau[\epsilon(t)] = |t|. \tag{1.6}$$

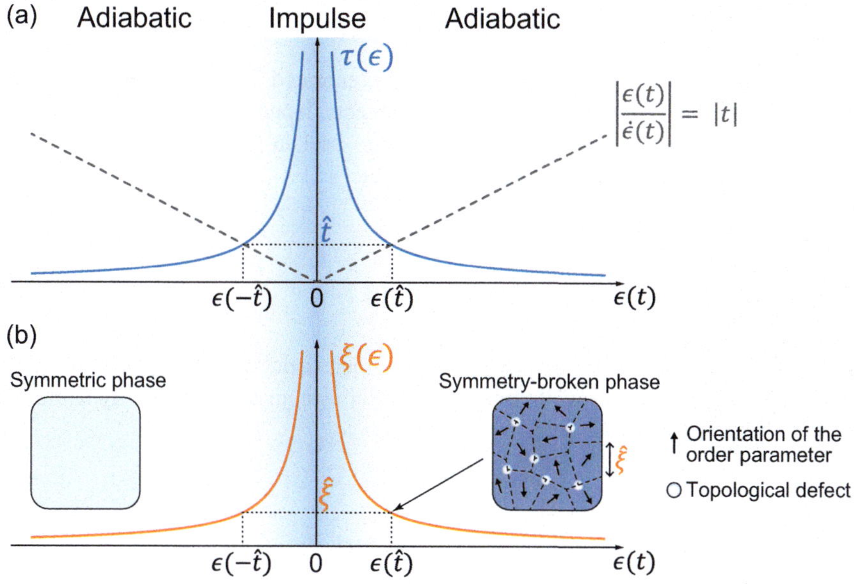

Fig. 1.13 Schematic representation of the adiabatic-impulse-adiabatic approximation predicted by the Kibble-Zurek mechanism. In the impulse regime between $\epsilon(-\hat{t})$ and $\epsilon(\hat{t})$, the system is effectively frozen due to the critical slowing down. **a** Relaxation time $\tau(\epsilon)$ and inverse transition rate $|\epsilon(t)/\dot{\epsilon}(t)|$. **b** Spatial correlation length $\xi(\epsilon)$. Inset: Schematics of the 2D system with 0D pointlike topological defects

Using Eq. 1.5 and $\epsilon(t) = t/t_q$, we can obtain the freeze-out time $\hat{t}$ (and $-\hat{t}$), which marks the impulse-adiabatic (adiabatic-impulse) crossover, and corresponding control parameter $\epsilon(\hat{t})$ as a function of quench time t_q or quench rate $\dot{\epsilon}$:

$$\hat{t} \propto t_q^{\frac{\nu z}{1+\nu z}} \propto \dot{\epsilon}^{-\frac{\nu z}{1+\nu z}}, \tag{1.7}$$

$$\epsilon(\hat{t}) \propto t_q^{-\frac{1}{1+\nu z}} \propto \dot{\epsilon}^{\frac{1}{1+\nu z}}. \tag{1.8}$$

The evolution of spatial correlation $\xi(\epsilon)$ also freezes between $\epsilon(-\hat{t})$ and $\epsilon(\hat{t})$. Therefore, when the system arrives at $\epsilon(\hat{t})$ after the symmetry-breaking transition, the order parameter becomes a mosaic of ordered domains whose average size $\hat{\xi}$ is estimated as

$$\hat{\xi} \equiv \xi[\epsilon(\hat{t})] \propto t_q^{\frac{\nu}{1+\nu z}} \propto \dot{\epsilon}^{-\frac{\nu}{1+\nu z}}. \tag{1.9}$$

Consequently, the fragment of the symmetric disordered phase remains at boundaries of symmetry-broken ordered domains as topological defects as illustrated in the inset of Fig. 1.13b. The key prediction of the KZM is the power-law scaling of the density P_D of d-dimensional topological defects in the D-dimensional system:

$$P_D \propto \frac{\hat{\xi}^d}{\hat{\xi}^D} \propto \left(\frac{1}{t_q}\right)^{\frac{(D-d)\nu}{1+\nu z}} \propto \dot{\epsilon}^{\frac{(D-d)\nu}{1+\nu z}}. \tag{1.10}$$

This indicates that more topological defects are generated with increasing the quench rate. It is noted that the type of topological defects depends on the topology of the manifold of degenerated states in the symmetry-broken phase, which is related to the number of components of the order parameter [219, 220].

1.3 Superconducting Vortex System

In this section, we outline the property of the superconducting vortex system, which is used in this work as a well-controlled model system to study the dynamics of externally driven many particles.

1.3.1 Classification of Superconductors

Superconductivity is the phenomenon characterized by exactly zero resistivity and perfect diamagnetism below a critical temperature T_c [224, 225]. It has been a major question since old times what the electrical resistance would be at very low temperatures. Some scientists thought that the resistance of conductors would become

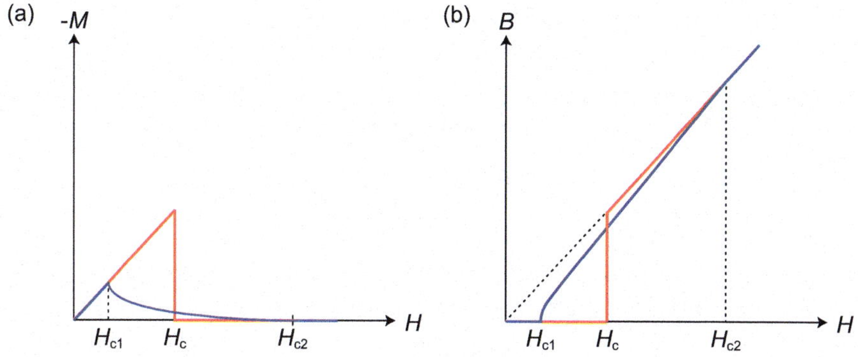

Fig. 1.14 Comparison of magnetic properties between type-I and type-II superconductors. **a** Magnetization. **b** Flux penetration behavior. Type-I superconductors (red) show the Meissner state ($H < H_c$) and the normal state ($H > H_c$). Type-II superconductors (blue) show the Meissner state ($H < H_{c1}$), the mixed state ($H_{c1} < H < H_{c2}$), and the normal state ($H > H_{c2}$)

infinitely large at absolute zero due to freezing of carriers, while others considered that the resistance would gradually decrease and continuously go to zero with decreasing temperature. In 1911, Kamerlingh Onnes measured the resistivity of mercury immersed in liquid helium which he had first liquefied in 1908. Contrary to most expectations, he found that the resistance abruptly vanished at 4.2 K. This was the discovery of superconductivity ("supraconductivity" as he preferred to call it) [226]. Another hallmark of superconductivity is perfect diamagnetism, discovered in 1933 by Meissner and Ochsenfeld. They found that the superconductor excludes an external magnetic field, and moreover, expels the initially penetrated field as it is cooled below T_c. This Meissner effect cannot be explained by perfect conductivity alone, which keeps trapping flux in.

Generally, superconductors are classified into two types, i.e., type-I and type-II, based on the response to the external field H. Figure 1.14 schematically shows comparisons of magnetization $M(H)$ (a) and flux penetration behavior $B(H)$ (b) between type-I and type-II superconductors, where $B(= \mu_0 H + M)$ is a magnetic flux density in the material and μ_0 is the magnetic constant. For type-I superconductors, as the external field is increased, Meissner state with perfect diamagnetism persists at a critical field H_c wherein the first-order transition from the superconducting state to the normal state occurs. Above H_c, superconductivity is destroyed and magnetic flux starts to penetrate the material. On the other hand, type-II superconductors have two critical fields H_{c1} and H_{c2}. Below the lower critical field H_{c1}, the material keeps Meissner state and above the upper critical field H_{c2}, the superconductivity breaks down as well as type-I superconductors. Between H_{c1} and H_{c2}, unlike the case of type-I superconductors, the Meissner state continuously changes into the so-called mixed state or the vortex state, where the magnetic field partially penetrates the superconductor in the form of an Abrikosov vortex [227].

1.3.2 Vortices in Type-II Superconductors

The Abrikosov vortex is a whirlpool of a supercurrent in which the flux is concentrated [228, 229]. Each vortex carries a quantum of flux

$$\Phi_0 = \frac{h}{2e} = 2.07 \times 10^{-15}\ \text{Wb}. \tag{1.11}$$

There are two characteristic length scales: the superconducting coherence length ξ and the London penetration depth λ. The coherence length is expressed as

$$\xi = \sqrt{\frac{\Phi_0}{2\pi \mu_0 H_{c2}}} \tag{1.12}$$

and characterizes the vortex core size. The vortex has a normal, non-superconducting core, which is separated from the surrounding superconducting region by the circulating supercurrent. The superconducting order parameter $|\Psi(r)|$ drops to zero towards the center ($r = 0$) of the vortex and a microscopic magnetic field $h(r)$ is distributed in the vortex core. The circulating supercurrent and $h(r)$ decay within the penetration depth λ, which determines the effective length of the vortex-vortex interaction. Figure 1.15 illustrates the variation of $|\Psi(r)|$ and $h(r)$ around the isolated vortex. Here, we sketch the straight vortex string whose degree of freedom of motion is two dimensions. When the sample thickness is larger than λ, the bending distortion and winding of the vortex occur along the field direction. Then the vortex system is regarded as the 3D system.

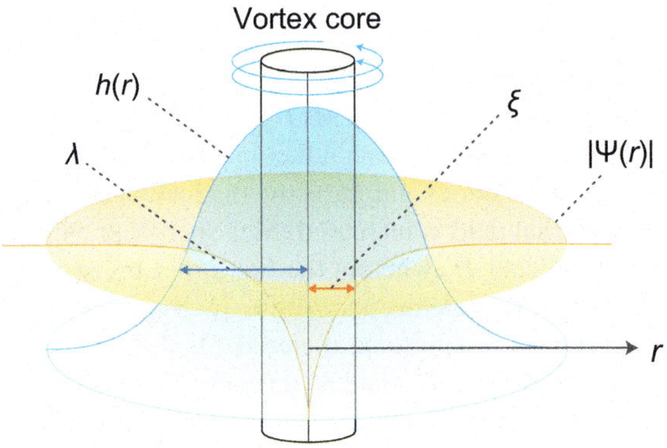

Fig. 1.15 Variations of the superconducting order parameter $|\Psi(r)|$ and the microscopic magnetic field $h(r)$ around the isolated vortex. $|\Psi(r)|$ and $h(r)$ decay within the length scales ξ and λ, respectively. Blue arrows represent the supercurrent circulating around the core

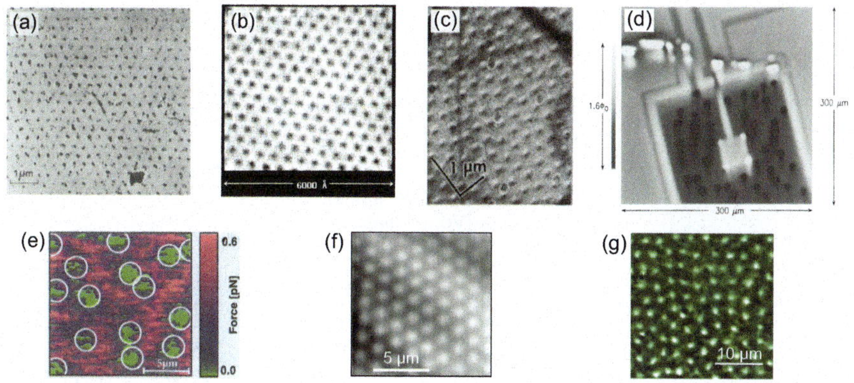

Fig. 1.16 Visualization of vortices by various imaging techniques. **a** Bitter decoration. **b** scanning tunneling microscopy. **c** Lorentz microscopy. **d** scanning SQUID microscopy. **e** magnetic force microscopy. **f** scanning Hall probe microscopy. **g** magneto-optical imaging. Adapted with permission from: **a** Ref. [231], Elsevier; **b** Ref. [236], American Physical Society; **c** Ref. [239], Springer Nature; **d** Ref. [244], AIP Publishing; **e** Ref. [246], American Physical Society; **f** Ref. [250], Springer Nature; **g** Ref. [252], IOP Publishing

Since the vortex-vortex interaction is repulsive, to minimize a free energy, vortices form the triangular lattice known as an Abrikosov lattice. The existence of vortex lattice was theoretically predicted by Abrikosov [227] and later experimentally demonstrated by various visualization techniques [230], such as Bitter decoration [231–235], scanning tunneling microscopy (STM) [236–238], Lorentz microscopy [239–242], scanning SQUID microscopy [243–245], magnetic force microscopy [246, 247], scanning Hall probe microscopy [248–251], and magneto-optical imaging [252], as representatively shown in Fig. 1.16. When vortices form the Abrikosov lattice (the perfect triangular lattice), its lattice spacing a_0 is easily calculated as

$$a_0 = \sqrt{\frac{2\Phi_0}{\sqrt{3}B}}. \tag{1.13}$$

The actual superconductor has random inhomogeneities or quenched disorder, such as impurities and defects, around which the material fails to become the superconducting state. Vortices with non-superconducting core are energetically more stable when arranged in normal regions, so the quenched disorder acts as the pinning center, that is, a source of the attractive force for vortices. Therefore, the vortex lattice is usually distorted from the perfect triangular lattice.

Figure 1.17b, c schematically illustrate the vortex configuration in equilibrium, where blue circles and red crosses indicate the vortices and the pinning centers, respectively. At low field B and temperature T in the presence of pinning, the perfect lattice loses translational order and deforms to a so-called Bragg glass, which has a quasi-long-range positional order and a long-range orientational order. In the Bragg

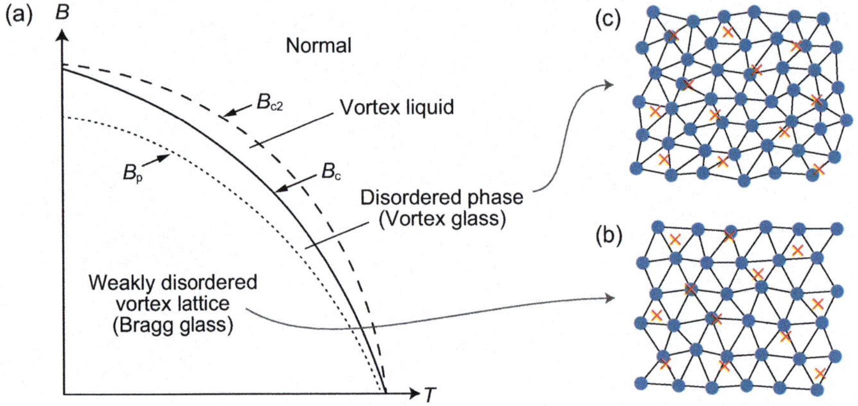

Fig. 1.17 **a** Typical equilibrium B–T phase diagram of vortices in the presence of pinning. The field B_c and B_{c2} mark the melting transition and the crossover from the vortex liquid to the normal state, respectively. B_p is a field at which the depinning current $I_d(B)$ takes a peak. **b**, **c** Schematic illustration of vortex arrangement. **b** Weakly disordered vortex lattice (Bragg glass). **c** Disordered phase (vortex glass). Blue circles and red crosses represent the vortices and the pinning centers, respectively

glass state or an ordered phase, the number of nearest neighbors is still six and topological defects are absent as shown in Fig. 1.17b. With an increase in B or T, the Bragg glass transforms into a disordered glassy phase called the vortex glass shown in Fig. 1.17c, in which many topological defects arise and the orientational order is lost. A transition from the ordered phase to the disordered phase is known as the order-disorder transition [253–257].

As B and T are further increased, the effect of thermal fluctuations is enhanced and the vortex solid melts into a vortex liquid, in which vortices can move freely. This transition is called a melting transition [258–263]. Figure 1.17a shows a typical B–T phase diagram of vortices in 3D superconductors. The field B_c and B_{c2} mark the melting transition and the crossover from the vortex liquid state to the normal state, respectively. B_p is a field at which the depinning current $I_d(B)$ takes a peak as discussed below.

1.3.3 Vortex Flow

We have explained the static property of vortices so far. In this section, we focus on the dynamics of vortices. When the external current is applied to the vortex system, each vortex feels a Lorentz-like force f per unit length:

$$\boldsymbol{f} = \boldsymbol{J} \times \boldsymbol{\Phi_0}, \tag{1.14}$$

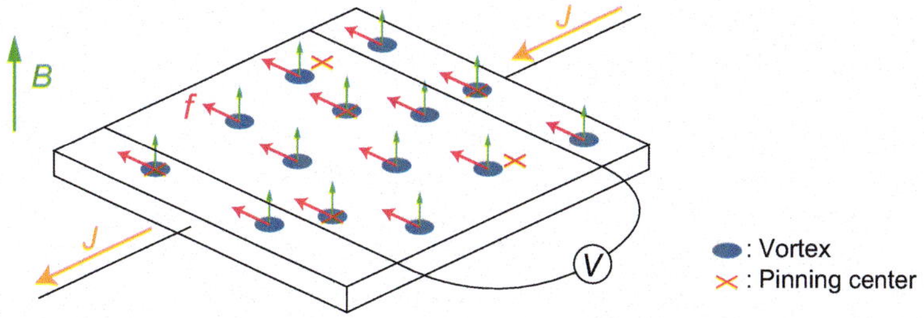

Fig. 1.18 Schematic illustration of vortices driven by the external current. Blue circles and red crosses indicate the vortices and the pinning centers, respectively. Green, yellow and magenta arrows represent the direction of the magnetic field $\boldsymbol{B}$, current $\boldsymbol{J}$, and the Lorentz-like driving force $\boldsymbol{f}$, respectively. The motion of vortices generates the voltage of Eq. 1.17

where $\boldsymbol{J}$ is a current density and $\boldsymbol{\Phi}_0$ is a vector with magnitude Φ_0 whose direction is parallel to the flux line. Using the formulation of the magnetic field $\boldsymbol{B} = n\boldsymbol{\Phi}_0$, where n is the density of vortices, the Lorentz-like force per unit area can be written as

$$\boldsymbol{F} = \boldsymbol{J} \times \boldsymbol{B}. \tag{1.15}$$

Thus, vortices move transverse to the applied current. The moving vortices with velocity $\boldsymbol{v}$ induce an electric field $\boldsymbol{E}$:

$$\boldsymbol{E} = \boldsymbol{B} \times \boldsymbol{v}, \tag{1.16}$$

which is parallel to $\boldsymbol{J}$, so that finite resistance arises even in the superconducting state. The voltage generated by the vortex motion between voltage contacts separated at l is expressed as

$$V = vBl. \tag{1.17}$$

Therefore, we can obtain the average velocity of vortices from the voltage measurements and estimate the dynamics of vortices. Figure 1.18 schematically illustrates the vortices (blue circles) driven by the dc current $\boldsymbol{J}$ (yellow arrows), which exerts the Lorentz-like driving force $\boldsymbol{f}$ (magenta arrows) on the vortices.

1.3.4 Peak Effect

As mentioned in Sect. 1.1.3, when the applied current exceeds the depinning current $I_{\rm d}$, pinned vortices undergo the depinning transition and start to flow. The value of $I_{\rm d}$ reflecting effective pinning force depends on the temperature and the magnetic field

B. Figure 1.19a illustrates a typical field dependence of I_d, where I_d decreases with B at low field. If vortices are trapped by pinning individually, the depinning current is determined only by the intrinsic pinning force and independent of B. However, the vortices in the stiff vortex lattice are hard to rearrange and hence cannot sit in the pinning centers one by one. Accordingly, the stiff vortex lattice is pinned collectively. With an increase in B, the amount of the Lorentz-like force acting on the lattice increases in proportion to B, while the total pinning force acting on the lattice does not increase so much. Consequently, I_d shows a decrease with increasing B. As B is further increased, however, I_d exhibits a rise and a peak at a certain field B_p. This phenomenon is known as a peak effect [264–273] and considered as a result of softening of the vortex lattice. In the peak-effect regime, the deformation of the soft vortex lattice allows more vortices to fit in the pinning centers, so pinning becomes more effective. A further increase in B leads to decreasing I_d followed by the melting transition at B_c. In the liquid phase, the vortices move freely due to the assist of thermal fluctuations, and hence I_d is zero.

Figure 1.19b–f show STM images of the vortex lattice as a function of B in the first row and corresponding Delauney triangulation in the second row, reported by Iavarone et al. [274]. Well below the onset of the peak-effect regime [Fig. 1.19b] and at the onset [Fig. 1.19c], vortices form the perfect lattice or the Bragg glass, indicating the ordered phase. Between the onset and the peak B_p [Fig. 1.19d], a small number of dislocations appear in the vortex lattice as clearly indicated by red triangles and blue squares in Delauney triangulation. In the peak-effect regime, the ordered phase and the disordered phase are coexistent. At the peak B_p [Fig. 1.19e], the order-disorder transition occurs. The vortex arrangement becomes disordered above B_p [Fig. 1.19f] as indicated by a large number of dislocations.

Recent studies [275–277] have shown that the vortex configuration seemingly disordered has a hidden order termed "Hyperuniformity" [278, 279]. The hyperuniform structure, which is characterized by a suppression of long-wavelength density fluctuations, is considered to appear in many places from structure of the Universe [280] to photoreceptor cells in avian retina [281]. The hyperuniformity is also expected to be observed near the critical point of the absorbing phase transition [282–288], which is of great interest but beyond the scope of this thesis.

1.3.5 RIT in the Vortex System

The vortex systems are suitable for studying nonequilibrium phase transitions experimentally for the following reasons: (i) we can apply the well-controlled driving force by the applied current (Eq. 1.15). (ii) we can readily control the density of vortices and mean intervortex spacing a_0 (Eq. 1.13) by changing the applied magnetic field. (iii) we can precisely detect the dynamics of vortices by measuring the voltage induced by their motion (Eq. 1.17). In fact, our group has reported the nonequilibrium depinning transition [45, 46, 49, 59] and the nonequilibrium RIT [59, 71, 75, 79], both of which belong to the DP universality class of absorbing phase transitions.

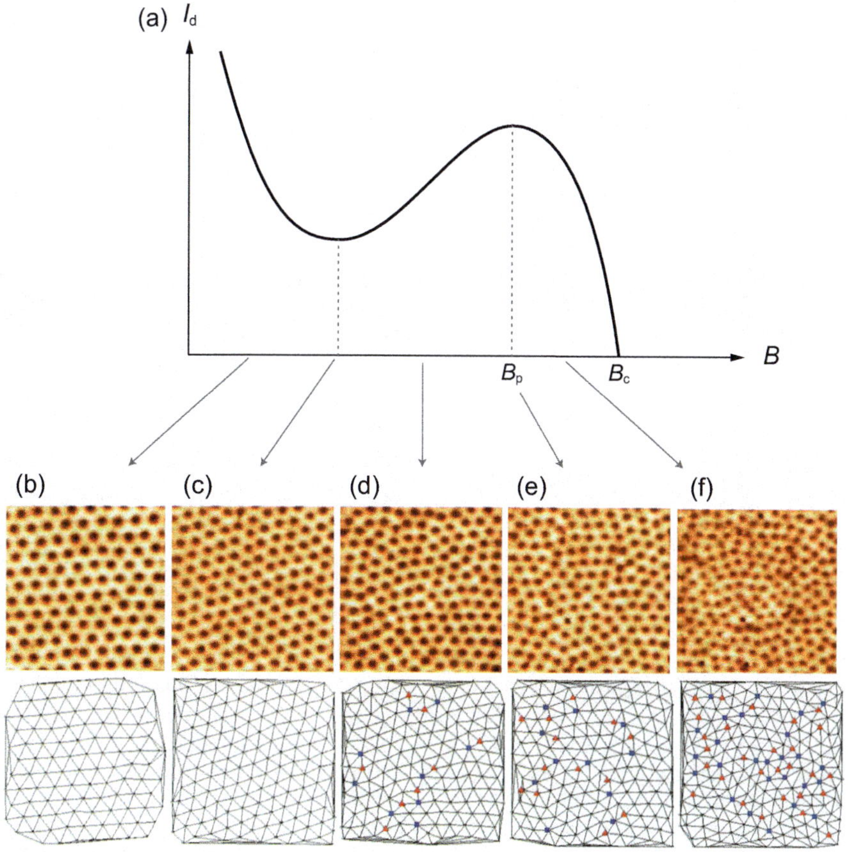

Fig. 1.19 Peak effect of the depinning current I_d as a function of the magnetic field B. **a** Schematic of the peak effect. B_p and B_c represent the peak field and the melting field, respectively. **b–f** STM images of static vortex configuration in the first row and corresponding Delauney triangulation in the second row. **b** Well below the onset of the peak-effect regime, **c** at the onset, **d** between the onset and the peak B_p, **e** at B_p, and **f** above B_p. Red triangles and blue squares in the Delauney triangulation indicate the dislocations with coordination number 5 and 7, respectively. The scanning area for STM images is 375×375 nm^2. **b–f** Reprinted figure with permission from Ref. [274]. Copyright 2008 by the American Physical Society

Here, let us briefly review the previous experiments of RIT in the vortex system caused by the global shearing force and the local shearing force.

In 2011, three years after first observation of RIT in colloidal suspensions, Okuma et al. conducted shearing experiments for vortices and proved that RIT is a universal phenomenon in sheared particlelike systems [59]. They used a Corbino-disk-shaped amorphous (a-)Mo_xGe_{1-x} film with random pinning as illustrated in the left bottom inset of Fig. 1.20a. By applying a radial ac square current I, the vortices rotate azimuthal direction periodically by the Lorentz-like force inversely proportional to

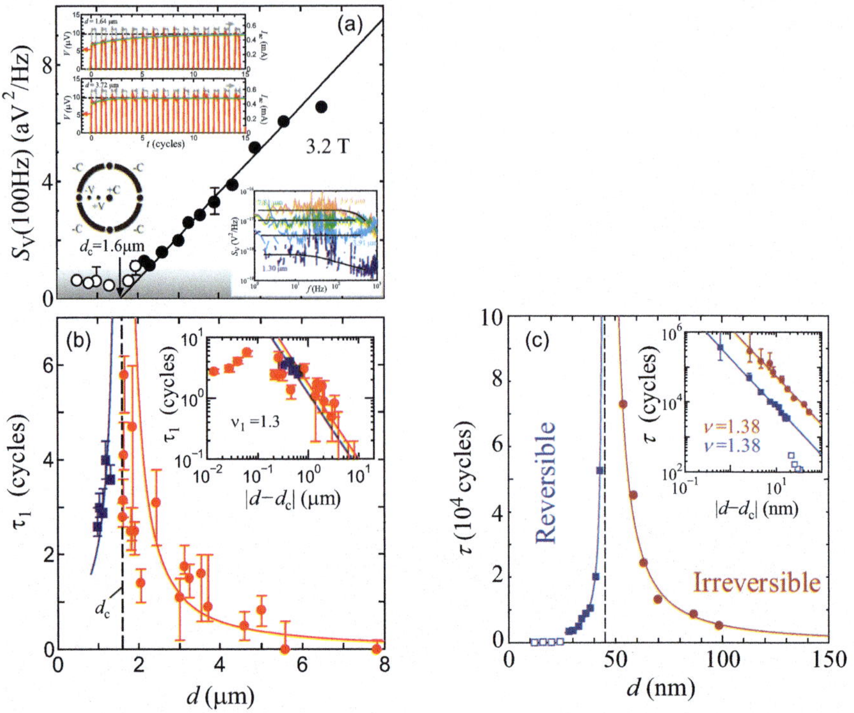

Fig. 1.20 **a** Noise Spectra S_V versus shear amplitude d [59]. Open (solid) circles represent S_V below (above) the background level. The arrow marks d_c. Inset: (Left top) Gray dots and a horizontal dashed line represent the ac current and amplitude of ac voltage in the steady state. Red lines represent $V(t)$ obtained for $d = 1.64$ μm and $d = 3.72$ μm. (Left bottom) Schematic illustration of electrical contact arrangement. **b**, **c** $\tau(d)$ for RIT under artificial global shear (**b**) [59] and random local shear (**c**) [75]. Blue squares and red circles are the data for the reversible state and the irreversible state, respectively. Full lines on both sides of RIT show the fits to $\tau \propto |d - d_c|^{-\nu}$ with $\nu = 1.3 \pm 0.3$ (**b**) and $\nu = 1.38 \pm 0.08$ (**c**). Inset of (**b**, **c**): The same data in the main panel against $|d - d_c|$ on a log-log scale. **a**, **b** Reprinted figure with permission from Ref. [59]. Copyright 2011 by the American Physical Society. **c** Reprinted from Ref. [75] by S. Maegochi et al. licensed under CC BY 4.0

the radius of rotation. This creates an artificial global shear for the vortices as shown in Fig. 1.21a, where the geometry of the Corbino disk is similar to that of the Couette cell used in colloidal experiments.

Figure 1.20a shows broadband voltage noise spectra S_V (steady-state flow noise of vortices), which are below background level (open circles) for $d < 2$ μm, while the substantial noise (solid circles) appears for large d. Here,

$$d = \frac{v^\infty}{2f} = \frac{V^\infty}{2Blf} \tag{1.18}$$

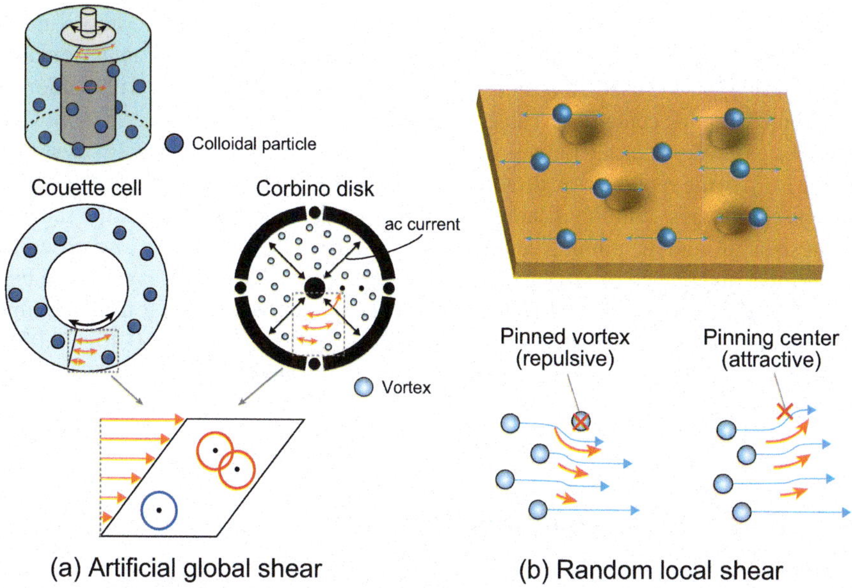

Fig. 1.21 Two types of shear in RIT experiments. **a** Artificial global shear in the colloidal experiment [53] and in the vortex experiment using the Corbino-disk geometry [59]. **b** Random local shear in vortex experiments using the strip-shaped sample [75]. Blue circles and red crosses represent the vortices and the pinning sites, respectively. The repulsive force and attractive force are exerted on the vortices from pinned vortices and pinning sites, respectively. The magnitude of the force decays with the distance from the pinning center. Blue arrows depict trajectories of the vortices and orange arrows represent the local shear around the random pinning sites

is a displacement of the vortex by the global shearing force, where $v^{\infty}(= V^{\infty}/Bl)$ is a steady-state average velocity of vortices calculated from Eq. 1.17 and $(1/2f)$ is a half period of the ac square current. The onset of S_V at $d_c \approx 1.6$ µm can be considered as the critical point of RIT above which the vortex motion is diffusive.

Okuma et al. also observed the random organization of periodically sheared vortices by measuring the transient voltage response. They found that vortices with disordered initial configuration gradually self-organize and become more mobile by the periodic shear as shown in the left top inset of Fig. 1.20a, where the voltage (average velocity of vortices) increases toward the steady-state value V^{∞} with increasing the number of shear cycles. Figure 1.20b shows that the relaxation time τ extracted from the voltage relaxation exhibits the power-law divergence $\tau \propto |d - d_c|^{-\nu}$ with $\nu = 1.3 \pm 0.3$, indicative of the nonequilibrium RIT. The value of the critical exponent $\nu = 1.3 \pm 0.3$ is slightly larger than $\nu = 1.1 \pm 0.3$ in the colloidal experiment [Fig. 1.6b] but is close to $\nu = 1.33 \pm 0.02$ in the two-dimensional colloidal simulation [Fig. 1.6a] and $\nu_{\parallel} = 1.295 \pm 0.006$ predicted for the DP universality class in 2D (Table 1.1). This result agrees with the fact that the vortex system studied here is regarded as the 2D system.

Since RIT was demonstrated under the artificial global shear, we have further investigated this transition with more general situations, that is, RIT under random local shear [75]. We used an ordinary strip-shaped a-Mo_xGe_{1-x} film with randomly distributed pinning centers that cause local shearing force around them. When moving vortices pass very near the pinned vortices or the unoccupied pinning sites as shown in Fig. 1.21b, the repulsive or attractive force is exerted on them, respectively, whose magnitude decays with the distance from pinning centers. This creates the random local shear around the pinning centers. For the random local shear, we also observed the critical divergence of $\tau(d)$ at $d_c \approx 45$ nm with $\nu = 1.38 \pm 0.08$ as shown in Fig. 1.20c. In addition, we obtained another critical exponent α in Eq. 1.2 from the time evolution of active irreversible vortices around d_c. The value of $\alpha = 0.40 \pm 0.05$ is close to $\alpha \approx 0.45$ for the 2D DP class. More details and development of the study will be provided in Chap. 3.

1.4 Purposes and Outline of This Thesis

In recent years, there has been a surge of interest in nonequilibrium phase transitions as shown before. Compared to equilibrium phase transitions, however, the study on the nonequilibrium phase transitions is still in its infancy and for further understanding of them, more exploration is needed from both theoretical and experimental sides. Given that observations of the DP critical behavior have only recently been reported in a limited number of experimental systems, whereas there are many theoretical models as mentioned in Sect. 1.1.2, the experimental work seems to be less advanced than the theoretical one. In view of this, we perform experimental studies on nonequilibrium phase transitions in externally driven many-particle systems by using the vortex system, which is suitable for studying the nonequilibrium phase transitions as explained in the previous section. In particular, we focus on the reversible-irreversible transition (RIT) induced by the ac drive, which has been found to exhibit DP criticality [53, 59, 63, 71, 72, 75, 79], and the dynamical ordering transition induced by the dc drive.

This thesis is organized as follows. In Chap. 2, we explain the overall experimental setup. In Chap. 3, RIT driven by particle (vortex) density as well as shear amplitude is studied. We also show two distinct flow regimes in the reversible phase. In Chap. 4, we explore current-induced dynamical ordering of dc driven vortices. We develop the experimental methods: two-step measurements and transverse mode-locking measurements. Dynamical phase diagram of dc driven vortices is presented together with convincing evidence of the moving smectic phase. In Chap. 5, we develop measurements of transverse current-voltage (force-velocity) characteristics and provide firm evidence of the second-order phase transition for the dynamical ordering transition. In Chap. 6, we demonstrate the Kibble-Zurek mechanism for the dynamical ordering transition. The summary of this thesis with an outlook is presented in Chap. 7.

References

1. Cross MC, Hohenberg PC (1993) Pattern formation outside of equilibrium. Rev Mod Phys 65:851
2. Whitesides GM, Ismagilov RF (1999) Complexity in chemistry. Science 284:89
3. Parrish JK, Edelstein-Keshet L (1999) Complexity, pattern, and evolutionary trade-offs in animal aggregation. Science 284:99
4. Werner BT (1999) Complexity in natural landform patterns. Science 284:102
5. Grzybowski BA, Stone HA, Whitesides GM (2000) Dynamic self-assembly of magnetized, millimetre-sized objects rotating at a liquid-air interface. Nature 405:1033
6. Whitesides GM, Grzybowski B (2002) Self-assembly at all scales. Science 295:2418
7. Aranson IS, Tsimring LS (2006) Patterns and collective behavior in granular media: theoretical concepts. Rev Mod Phys 78:641
8. Ramaswamy S (2010) The mechanics and statistics of active matter. Annu Rev Condens Matter Phys 1:323
9. Vicsek T, Zafeiris A (2012) Collective motion. Phys Rep 517:71
10. Marchetti MC, Joanny JF, Ramaswamy S, Liverpool TB, Prost J, Rao M, Simha RA (2013) Hydrodynamics of soft active matter. Rev Mod Phys 85:1143
11. Rubenstein M, Cornejo A, Nagpal R (2014) Programmable self-assembly in a thousand-robot swarm. Science 345:795
12. Bechinger C, Di Leonardo R, Löwen H, Reichhardt C, Volpe G, Volpe G (2016) Active particles in complex and crowded environments. Rev Mod Phys 88:045006
13. Fodor É, Marchetti MC (2018) The statistical physics of active matter: from self-catalytic colloids to living cells. Physica A 504:106
14. Nishiguchi D, Aranson IS, Snezhko A, Sokolov A (2018) Engineering bacterial vortex lattice via direct laser lithography. Nat Commun 9:4486
15. Gompper G, Winkler RG, Speck T, Solon A, Nardini C, Peruani F, Löwen H, Golestanian R, Kaupp UB, Alvarez L, Kiørboe T, Lauga E, Poon WCK, DeSimone A, Muiños-Landin S, Fischer A, Söker NA, Cichos F, Kapral R, Gaspard P, Ripoll M, Sagues F, Doostmohammadi A, Yeomans JM, Aranson IS, Bechinger C, Stark H, Hemelrijk CK, Nedelec FJ, Sarkar T, Aryaksama T, Lacroix M, Duclos G, Yashunsky V, Silberzan P, Arroyo M, Kale S (2020) The 2020 motile active matter roadmap. J Phys: Condens Matter 32:193001
16. Sciortino A, Bausch AR (2021) Pattern formation and polarity sorting of driven actin filaments on lipid membranes. Proc Natl Acad Sci USA 118:e2017047118
17. Buhl J, Sumpter DJT, Couzin ID, Hale JJ, Despland E, Miller ER, Simpson SJ (2006) From disorder to order in marching locusts. Science 312:1402
18. Cavagna A, Cimarelli A, Giardina I, Parisi G, Santagati R, Stefanini F, Viale M (2010) Scale-free correlations in starling flocks. Proc Natl Acad Sci USA 107:11865
19. Bialek W, Cavagna A, Giardina I, Mora T, Silvestri E, Viale M, Walczak AM (2012) Statistical mechanics for natural flocks of birds. Proc Natl Acad Sci USA 109:4786
20. Bricard A, Caussin J-B, Desreumaux N, Dauchot O, Bartolo D (2013) Emergence of macroscopic directed motion in populations of motile colloids. Nature 503:95
21. Morin A, Desreumaux N, Caussin J-B, Bartolo D (2017) Distortion and destruction of colloidal flocks in disordered environments. Nat Phys 13:63
22. Cates ME, Tailleur J (2015) Motility-induced phase separation. Annu Rev Condens Matter Phys 6:219
23. Fily Y, Marchetti MC (2012) Athermal phase separation of self-propelled particles with no alignment. Phys Rev Lett 108:235702
24. Buttinoni I, Bialké J, Kümmel F, Löwen H, Bechinger C, Speck T (2013) Dynamical clustering and phase separation in suspensions of self-propelled colloidal particles. Phys Rev Lett 110:238301
25. Digregorio P, Levis D, Suma A, Cugliandolo LF, Gonnella G, Pagonabarraga I (2018) Full phase diagram of active Brownian disks: from melting to motility-induced phase separation. Phys Rev Lett 121:098003

26. Omar AK, Klymko K, GrandPre T, Geissler PL (2021) Phase diagram of active Brownian spheres: crystallization and the metastability of motility-induced phase separation. Phys Rev Lett 126:188002
27. Adachi K, Takasan K, Kawaguchi K (2022) Activity-induced phase transition in a quantum many-body system. Phys Rev Res 4:013194
28. Reichhardt C, Reichhardt CJO (2017) Depinning and nonequilibrium dynamic phases of particle assemblies driven over random and ordered substrates: a review. Rep Prog Phys 80:026501
29. Fisher DS (1998) Collective transport in random media: from superconductors to earthquakes. Phys Rep 301:113
30. Parker ID, Zettl A (1992) Charge-density-wave pinning and metastable-state dynamics in $NbSe_3$. Phys Rev B 45:3260
31. Duruöz CI, Clarke RM, Marcus CM, Harris JS Jr (1995) Conduction threshold, switching, and hysteresis in quantum dot arrays. Phys Rev Lett 74:3237
32. Reichhardt C, Olson CJ (2002) Colloidal dynamics on disordered substrates. Phys Rev Lett 89:078301
33. Maeda A, Inoue Y, Kitano H, Savel'ev S, Okayasu S, Tsukada I, Nori F (2005) Nanoscale friction: kinetic friction of magnetic flux quanta and charge density waves. Phys Rev Lett 94:077001
34. Kolton AB, Rosso A, Giamarchi T, Krauth W (2006) Dynamics below the depinning threshold in disordered elastic systems. Phys Rev Lett 97:057001
35. Pertsinidis A, Ling XS (2008) Statics and dynamics of 2D colloidal crystals in a random pinning potential. Phys Rev Lett 100:028303
36. Reichhardt C, Reichhardt CJO (2009) Random organization and plastic depinning. Phys Rev Lett 103:168301
37. Burrowes C, Mihai AP, Ravelosona D, Kim J-V, Chappert C, Vila L, Marty A, Samson Y, Garcia-Sanchez F, Buda-Prejbeanu LD, Tudosa I, Fullerton EE, Attané J-P (2010) Non-adiabatic spin-torques in narrow magnetic domain walls. Nat Phys 6:17
38. Tamura K, Ozawa T, Bando Y, Kawamoto T, Mori T (2010) Voltage oscillation associated with nonlinear conductivity in the organic conductor α-$(BEDT\text{-}TTF)_2I_3$. J Appl Phys 107:103716
39. Fily Y, Olive E, Scala ND, Soret JC (2010) Critical behavior of plastic depinning of vortex lattices in two dimensions: molecular dynamics simulations. Phys Rev B 82:134519
40. Daroca DP, Lozano GS, Pasquini G, Bekeris V (2010) Depinning and dynamics of ac driven vortex lattices in random media. Phys Rev B 81:184520
41. Pi U-H, Cho Y-J, Bae J-Y, Lee S-C, Seo S, Kim W, Moon J-H, Lee K-J, Lee H-W (2011) Static and dynamic depinning processes of a magnetic domain wall from a pinning potential. Phys Rev B 84:024426
42. Daroca DP, Pasquini G, Lozano GS, Bekeris V (2011) Dynamics of superconducting vortices driven by oscillatory forces in the plastic-flow regime. Phys Rev B 84:012508
43. Tierno P (2012) Depinning and collective dynamics of magnetically driven colloidal monolayers. Phys Rev Lett 109:198304
44. Shaw G, Mandal P, Banerjee SS, Niazi A, Rastogi AK, Sood AK, Ramakrishnan S, Grover AK (2012) Critical behavior at depinning of driven disordered vortex matter in $2H$-NbS_2. Phys Rev B 85:174517
45. Okuma S, Motohashi A (2012) Critical behavior associated with transient dynamics near the depinning transition. New J Phys 14:123021
46. Kawamura Y, Moriya S, Ienaga K, Kaneko S, Okuma S (2017) Nonequilibrium depinning transition of ac driven vortices with random pinning. New J Phys 19:093001
47. Brazda T, Silva A, Manini N, Vanossi A, Guerra R, Tosatti E, Bechinger C (2018) Experimental observation of the Aubry transition in two-dimensional colloidal monolayers. Phys Rev X 8:011050
48. Blay ML, Adda-Bedia M, Bartolo D (2020) Emergence of scale-free smectic rivers and critical depinning in emulsions driven through disorder. Proc Natl Acad Sci USA 117:13914

49. Kaji T, Maegochi S, Ienaga K, Kaneko S, Okuma S (2022) Critical behavior of nonequilibrium depinning transitions for vortices driven by current and vortex density. Sci Rep 12:1542
50. Reichhardt C, Regev I, Dahmen K, Okuma S, Reichhardt CJO (2023) Reversible to irreversible transitions in periodic driven many-body systems and future directions for classical and quantum systems. Phys Rev Res 5:021001
51. Pine DJ, Gollub JP, Brady JF, Leshansky AM (2005) Chaos and threshold for irreversibility in sheared suspensions. Nature 438:997
52. Gollub J, Pine D (2006) Microscopic irreversibility and chaos. Phys Today 59(8):8
53. Corté L, Chaikin PM, Gollub JP, Pine DJ (2008) Random organization in periodically driven systems. Nat Phys 4:420
54. Mangan N, Reichhardt C, Reichhardt CJO (2008) Reversible to irreversible flow transition in periodically driven vortices. Phys Rev Lett 100:187002
55. Düring G, Bartolo D, Kurchan J (2009) Irreversibility and self-organization in hydrodynamic echo experiments. Phys Rev E 79:030101(R)
56. Menon GI, Ramaswamy S (2009) Universality class of the reversible-irreversible transition in sheared suspensions. Phys Rev E 79:061108
57. Guasto JS, Ross AS, Gollub JP (2010) Hydrodynamic irreversibility in particle suspensions with nonuniform strain. Phys Rev E 81:061401
58. Zhang W, Zhou W, Luo M (2010) Irreversibility of two-dimensional vortex systems with random pinning. Phys Lett A 374:3666
59. Okuma S, Tsugawa Y, Motohashi A (2011) Transition from reversible to irreversible flow: absorbing and depinning transitions in a sheared-vortex system. Phys Rev B 83:012503
60. Slotterback S, Mailman M, Ronaszegi K, van Hecke M, Girvan M, Losert W (2012) Onset of irreversibility in cyclic shear of granular packings. Phys Rev E 85:021309
61. Schreck CF, Hoy RS, Shattuck MD, O'Hern CS (2013) Particle-scale reversibility in athermal particulate media below jamming. Phys Rev E 88:052205
62. Regev I, Lookman T, Reichhardt C (2013) Onset of irreversibility and chaos in amorphous solids under periodic shear. Phys Rev E 88:062401
63. Nagamanasa KH, Gokhale S, Sood AK, Ganapathy R (2014) Experimental signatures of a nonequilibrium phase transition governing the yielding of a soft glass. Phys Rev E 89:062308
64. Keim NC, Arratia PE (2014) Mechanical and microscopic properties of the reversible plastic regime in a 2D jammed material. Phys Rev Lett 112:028302
65. Jeanneret R, Bartolo D (2014) Geometrically protected reversibility in hydrodynamic Loschmidt-echo experiments. Nat Commun 5:3474
66. Regev I, Weber J, Reichhardt C, Dahmen KA, Lookman T (2015) Reversibility and criticality in amorphous solids. Nat Commun 6:8805
67. Pfeifer JC, Bischoff T, Ehlers G, Eckhardt B (2015) Order-disorder transitions in a sheared many-body system. Phys Rev E 92:062208
68. Royer JR, Chaikin PM (2015) Precisely cyclic sand: self-organization of periodically sheared frictional grains. Proc Natl Acad Sci USA 112:49
69. Pham P, Butler JE, Metzger B (2016) Origin of critical strain amplitude in periodically sheared suspensions. Phys Rev Fluids 1:022201(R)
70. Jana PK, Alava MJ, Zapperi S (2017) Irreversibility transition of colloidal polycrystals under cyclic deformation. Sci Rep 7:45550
71. Dobroka M, Kawamura Y, Ienaga K, Kaneko S, Okuma S (2017) Memory formation and evolution of the vortex configuration associated with random organization. New J Phys 19:053023
72. Ni X, Zhang H, Liarte DB, McFaul LW, Dahmen KA, Sethna JP, Greer JR (2019) Yield precursor dislocation avalanches in small crystals: the irreversibility transition. Phys Rev Lett 123:035501
73. Regev I, Reichhardt C, Reichhardt CJO (2019) Noise spectra in the reversible-irreversible transition in amorphous solids under oscillatory driving. Model Simul Mater Sci Eng 27:084004
74. Brown BL, Reichhardt C, Reichhardt CJO (2019) Reversible to irreversible transitions in periodically driven skyrmion systems. New J Phys 21:013001

75. Maegochi S, Ienaga K, Kaneko S, Okuma S (2019) Critical behavior near the reversible-irreversible transition in periodically driven vortices under random local shear. Sci Rep 9:16447
76. Nagasawa K, Miyazaki K, Kawasaki T (2019) Classification of the reversible-irreversible transitions in particle trajectories across the jamming transition point. Soft Matter 15:7557
77. Mukherji S, Kandula N, Sood AK, Ganapathy R (2019) Strength of mechanical memories is maximal at the yield point of a soft glass. Phys Rev Lett 122:158001
78. Das P, Vinutha HA, Sastry S (2020) Unified phase diagram of reversible-irreversible, jamming, and yielding transitions in cyclically sheared soft-sphere packings. Proc Natl Acad Sci USA 117:10203
79. Miyagawa K, Maegochi S, Ienaga K, Kaneko S, Okuma S (2021) Effects of the velocity on the reversible-irreversible transition in a periodically sheared vortex system. J Phys: Conf Ser 1975:012002
80. Frenkel D (2008) Ordered chaos. Nat Phys 4:345
81. Milz L, Schmiedeberg M (2013) Connecting the random organization transition and jamming within a unifying model system. Phys Rev E 88:062308
82. Zhou C, Reichhardt CJO, Reichhardt C, Beyerlein I (2014) Random organization in periodically driven gliding dislocations. Phys Lett A 378:1675
83. Hexner D, Chaikin PM, Levine D (2017) Enhanced hyperuniformity from random reorganization. Proc Natl Acad Sci USA 114:4294
84. Ness C, Mari R, Cates ME (2018) Shaken and stirred: random organization reduces viscosity and dissipation in granular suspensions. Sci Adv 4:eaar3296
85. Lei Q-L, Ni R (2019) Hydrodynamics of random-organizing hyperuniform fluids. Proc Natl Acad Sci USA 116:22983
86. Martiniani S, Chaikin PM, Levine D (2019) Quantifying hidden order out of equilibrium. Phys Rev X 9:011031
87. Zheng Y, Parmar ADS, Ciamarra MP (2021) Hidden order beyond hyperuniformity in critical absorbing states. Phys Rev Lett 126:118003
88. Wilken S, Guerra RE, Levine D, Chaikin PM (2021) Random close packing as a dynamical phase transition. Phys Rev Lett 127:038002
89. Reynolds O (1883) An experimental investigation of the circumstances which determine whether the motion of water shall be direct or sinuous, and of the law of resistance in parallel channels. Phil Trans R Soc Lond A 174:935
90. Hinrichsen H (2000) Non-equilibrium critical phenomena and phase transitions into absorbing states. Adv Phys 49:815
91. Ódor G (2004) Universality classes in nonequilibrium lattice systems. Rev Mod Phys 76:663
92. Lübeck S (2004) Universal scaling behavior of non-equilibrium phase transitions. Int J Mod Phys B 18:3977
93. Henkel M, Hinrichsen H, Lübeck S (2008) Non-equilibrium phase transitions: vol. 1: absorbing phase transitions. Springer, Dordrecht
94. Livi R, Politi P (2017) Nonequilibrium statistical physics: a modern perspective. Cambridge University Press, Cambridge
95. Broadbent SR, Hammersley JM (1957) Percolation processes: i. crystals and mazes. Pros Camb Phil Soc 53:629
96. Arndt PF, Dahmen SR, Hinrichsen H (2001) Directed percolation, fractal roots and the Lee-Yang theorem. Physica A 295:128
97. Essam JW (1980) Percolation theory. Rep Prog Phys 43:833
98. Muñoz MA (2018) Colloquium: criticality and dynamical scaling in living systems. Rev Mod Phys 90:031001
99. Mollison D (1977) Spatial contact models for ecological and epidemic spread. J R Stat Soc B 39:283
100. Pomeau Y (1986) Front motion, metastability and subcritical bifurcations in hydrodynamics. Physica D 23:3

101. Grinstein G, Lai Z-W, Browne DA (1989) Critical phenomena in a nonequilibrium model of heterogeneous catalysis. Phys Rev A 40:4820(R)
102. Jensen I, Fogedby HC, Dickman R (1990) Critical exponents for an irreversible surface reaction model. Phys Rev A 41:3411(R)
103. Voigt CA, Ziff RM (1997) Epidemic analysis of the second-order transition in the Ziff-Gulari-Barshad surface-reaction model. Phys Rev E 56:R6241(R)
104. Schulman LS, Seiden PE (1986) Percolation and galaxies. Science 233:425
105. Seiden PE, Schulman LS (1990) Percolation model of galactic structure. Adv Phys 39:1
106. Hinrichsen H, Jiménez-Dalmaroni A, Rozov Y, Domany E (1999) Flowing sand: a physical realization of directed percolation. Phys Rev Lett 83:4999
107. Grassberger P (1997) Nonlinearities in complex systems. In: Puri S et al (ed) Proceedings of the 1995 Shimla conference on complex systems. Narosa Publishing, New Dehli
108. Takeuchi KA, Kuroda M, Chaté H, Sano M (2007) Directed percolation criticality in turbulent liquid crystals. Phys Rev Lett 99:234503
109. Takeuchi KA, Kuroda M, Chaté H, Sano M (2009) Experimental realization of directed percolation criticality in turbulent liquid crystals. Phys Rev E 80:051116
110. Takeuchi KA (2014) Experimental approaches to universal out-of-equilibrium scaling laws: turbulent liquid crystal and other developments. J Stat Mech P01006
111. Sano M, Tamai K (2016) A universal transition to turbulence in channel flow. Nat Phys 12:249
112. Lemoult G, Shi L, Avila K, Jalikop SV, Avila M, Hof B (2016) Directed percolation phase transition to sustained turbulence in Couette flow. Nat Phys 12:254
113. Helmrich S, Arias A, Lochead G, Wintermantel TM, Buchhold M, Diehl S, Whitlock S (2020) Signatures of self-organized criticality in an ultracold atomic gas. Nature 577:481
114. Chantelot P, Lohse D (2021) Leidenfrost effect as a directed percolation phase transition. Phys Rev Lett 127:124501
115. Reichhardt C, Olson CJ, Grønbech-Jensen N, Nori F (2001) Moving Wigner glasses and smectics: dynamics of disordered Wigner crystals. Phys Rev Lett 86:4354
116. Danneau R, Ayari A, Rideau D, Requardt H, Lorenzo JE, Ortega L, Monceau P, Currat R, Grübel G (2002) Motional ordering of a charge-density wave in the sliding state. Phys Rev Lett 89:106404
117. Chen J, Cao Y, Jiao Z (2004) Dynamics of two-dimensional colloids on a disordered substrate. Phys Rev E 69:041403
118. Reichhardt C, Reichhardt CJO, Milošević MV (2022) Statics and dynamics of skyrmions interacting with disorder and nanostructures. Rev Mod Phys 94:035005
119. Shi A-C, Berlinsky AJ (1991) Pinning and I-V characteristics of a two-dimensional defective flux-line lattice. Phys Rev Lett 67:1926
120. Koshelev AE, Vinokur VM (1994) Dynamic melting of the vortex lattice. Phys Rev Lett 73:3580
121. Giamarchi T, Le Doussal P (1996) Moving glass phase of driven lattices. Phys Rev Lett 76:3408
122. Balents L, Marchetti MC, Radzihovsky L (1998) Nonequilibrium steady states of driven periodic media. Phys Rev B 57:7705
123. Le Doussal P, Giamarchi T (1998) Moving glass theory of driven lattices with disorder. Phys Rev B 57:11356
124. Moon K, Scalettar RT, Zimányi GT (1996) Dynamical phases of driven vortex systems. Phys Rev Lett 77:2778
125. Olson CJ, Reichhardt C, Nori F (1998) Nonequilibrium dynamic phase diagram for vortex lattices. Phys Rev Lett 81:3757
126. Kolton AB, Domínguez D, Grønbech-Jensen N (1999) Hall noise and transverse freezing in driven vortex lattices. Phys Rev Lett 83:3061
127. Fangohr H, Cox DJ, de Groot PAJ (2001) Vortex dynamics in two-dimensional systems at high driving forces. Phys Rev B 64:064505
128. Ryu S, Hellerqvist M, Doniach S, Kapitulnik A, Stroud D (1996) Dynamical phase transition in a driven disordered vortex lattice. Phys Rev Lett 77:5114

129. Miguel M-C, Zapperi S (2003) Tearing transition and plastic flow in superconducting thin films. Nat Mater 2:477
130. Bhattacharya S, Higgins MJ (1993) Dynamics of a disordered flux line lattice. Phys Rev Lett 70:2617
131. Yaron U, Gammel PL, Huse DA, Kleiman RN, Oglesby CS, Bucher E, Batlogg B, Bishop DJ, Mortensen K, Clausen KN, Bolle CA, De La Cruz F (1994) Neutron diffraction studies of flowing and pinned magnetic flux lattices in 2 H-$NbSe_2$. Phys Rev Lett 73:2748
132. Yaron U, Gammel PL, Huse DA, Kleiman RN, Oglesby CS, Bucher E, Batlogg B, Bishop DJ, Mortensen K, Clausen KN (1995) Structural evidence for a two-step process in the depinning of the superconducting flux-line lattice. Nature 376:753
133. Marley AC, Higgins MJ, Bhattacharya S (1995) Flux flow noise and dynamical transition in a flux line lattice. Phys Rev Lett 74:3029
134. Hellerqvist MC, Ephron D, White WR, Beasley MR, Kapitulnik A (1996) Vortex dynamics in two-dimensional amorphous $Mo_{77}Ge_{23}$ films. Phys Rev Lett 76:4022
135. Pardo F, De La Cruz F, Gammel PL, Oglesby CS, Bucher E, Batlogg B, Bishop DJ (1997) Topological defects in the flux-line lattice and their relationship to the critical current of a Type-II superconductor. Phys Rev Lett 78:4633
136. Troyanovski AM, Aarts J, Kes PH (1999) Collective and plastic vortex motion in superconductors at high flux densities. Nature 399:665
137. Xiao ZL, Andrei EY, Higgins MJ (1999) Flow induced organization and memory of a vortex lattice. Phys Rev Lett 83:1664
138. Xiao ZL, Andrei EY, Shuk P, Greenblatt M (2000) Equilibration and dynamic phase transitions of a driven vortex lattice. Phys Rev Lett 85:3265
139. Togawa Y, Abiru R, Iwaya K, Kitano H, Maeda A (2000) Direct observation of the washboard noise of a driven vortex lattice in a high-temperature superconductor, $Bi_2Sr_2CaCu_2O_y$. Phys Rev Lett 85:3716
140. Fausti D, Tobey RI, Dean N, Kaiser S, Dienst A, Hoffmann MC, Pyon S, Takayama T, Takagi H, Cavalleri A (2011) Light-induced superconductivity in a stripe-ordered cuprate. Science 331:189
141. Tsuji N, Eckstein M, Werner P (2013) Nonthermal antiferromagnetic order and nonequilibrium criticality in the Hubbard model. Phys Rev Lett 110:136404
142. Stojchevska L, Vaskivskyi I, Mertelj T, Kusar P, Svetin D, Brazovskii S, Mihailovic D (2014) Ultrafast switching to a stable hidden quantum state in an electronic crystal. Science 344:177
143. Han T-RT, Zhou F, Malliakas CD, Duxbury PM, Mahanti SD, Kanatzidis MG, Ruan C-Y (2015) Exploration of metastability and hidden phases in correlated electron crystals visualized by femtosecond optical doping and electron crystallography. Sci Adv 1:e1400173
144. Oike H, Kagawa F, Ogawa N, Ueda A, Mori H, Kawasaki M, Tokura Y (2015) Phase-change memory function of correlated electrons in organic conductors. Phys Rev B 91:041101(R)
145. Zhang J, Tan X, Liu M, Teitelbaum SW, Post KW, Jin F, Nelson KA, Basov DN, Wu W, Averitt RD (2016) Cooperative photoinduced metastable phase control in strained manganite films. Nat Mater 15:956
146. Oike H, Kikkawa A, Kanazawa N, Taguchi Y, Kawasaki M, Tokura Y, Kagawa F (2016) Interplay between topological and thermodynamic stability in a metastable magnetic skyrmion lattice. Nat Phys 12:62
147. Kagawa F, Oike H (2017) Quenching of charge and spin degrees of freedom in condensed matter. Adv Mater 29:1601979
148. Berruto G, Madan I, Murooka Y, Vanacore GM, Pomarico E, Rajeswari J, Lamb R, Huang P, Kruchkov AJ, Togawa Y, LaGrange T, McGrouther D, Rønnow HM, Carbone F (2018) Laser-induced skyrmion writing and erasing in an ultrafast cryo-Lorentz transmission electron microscope. Phys Rev Lett 120:117201
149. Oike H, Kamitani M, Tokura Y, Kagawa F (2018) Kinetic approach to superconductivity hidden behind a competing order. Sci Adv 4:eaau3489
150. Nova TF, Disa AS, Fechner M, Cavalleri A (2019) Metastable ferroelectricity in optically strained $SrTiO_3$. Science 364:1075

151. Li X, Qiu T, Zhang J, Baldini E, Lu J, Rappe AM, Nelson KA (2019) Terahertz field-induced ferroelectricity in quantum paraelectric $SrTiO_3$. Science 364:1079
152. Zong A, Kogar A, Bie Y-Q, Rohwer T, Lee C, Baldini E, Ergeçen E, Yilmaz MB, Freelon B, Sie EJ, Zhou H, Straquadine J, Walmsley P, Dolgirev PE, Rozhkov AV, Fisher IR, Jarillo-Herrero P, Fine BV, Gedik N (2019) Evidence for topological defects in a photoinduced phase transition. Nat Phys 15:27
153. Kogar A, Zong A, Dolgirev PE, Shen X, Straquadine J, Bie Y-Q, Wang X, Rohwer T, Tung I-C, Yang Y, Li R, Yang J, Weathersby S, Park S, Kozina ME, Sie EJ, Wen H, Jarillo-Herrero P, Fisher IR, Wang X, Gedik N (2020) Light-induced charge density wave in $LaTe_3$. Nat Phys 16:159
154. Sun Z, Millis AJ (2020) Transient trapping into metastable states in systems with competing orders. Phys Rev X 10:021028
155. Zhou F, Williams J, Sun S, Malliakas CD, Kanatzidis MG, Kemper AF, Ruan C-Y (2021) Nonequilibrium dynamics of spontaneous symmetry breaking into a hidden state of charge-density wave. Nat Commun 12:566
156. Maklar J, Windsor YW, Nicholson CW, Puppin M, Walmsley P, Esposito V, Porer M, Rittmann J, Leuenberger D, Kubli M, Savoini M, Abreu E, Johnson SL, Beaud P, Ingold G, Staub U, Fisher IR, Ernstorfer R, Wolf M, Rettig L (2021) Nonequilibrium charge-density-wave order beyond the thermal limit. Nat Commun 12:2499
157. Matsuura K, Oike H, Kocsis V, Sato T, Tomioka Y, Kaneko Y, Nakamura M, Taguchi Y, Kawasaki M, Tokura Y, Kagawa F (2021) Kinetic pathway facilitated by a phase competition to achieve a metastable electronic phase. Phys Rev B 103:L041106
158. Basov DN, Averitt RD, Hsieh D (2017) Towards properties on demand in quantum materials. Nat Mater 16:1077
159. Keim NC, Paulsen JD, Zeravcic Z, Sastry S, Nagel SR (2019) Memory formation in matter. Rev Mod Phys 91:035002
160. Goldstein M (1969) Viscous liquids and the glass transition: a potential energy barrier picture. J Chem Phys 51:3728
161. Stillinger FH, Weber TA (1984) Packing structures and transitions in liquids and solids. Science 225:983
162. Stillinger FH (1995) A topographic view of supercooled liquids and glass formation. Science 267:1935
163. Becker OM, Karplus M (1997) The topology of multidimensional potential energy surfaces: theory and application to peptide structure and kinetics. J Chem Phys 106:1495
164. Wales DJ, Miller MA, Walsh TR (1998) Archetypal energy landscapes. Nature 394:758
165. Debenedetti PG, Stillinger FH, Truskett TM, Roberts CJ (1999) The equation of state of an energy landscape. J Phys Chem B 103:7390
166. Brooks CL III, Onuchic JN, Wales DJ (2001) Taking a walk on a landscape. Science 293:612
167. Wang C, Stratt RM (2007) Global perspectives on the energy landscapes of liquids, supercooled liquids, and glassy systems: the potential energy landscape ensemble. J Chem Phys 127:224503
168. Wang C, Stratt RM (2007) Global perspectives on the energy landscapes of liquids, supercooled liquids, and glassy systems: geodesic pathways through the potential energy landscape. J Chem Phys 127:224504
169. Raza Z, Alling B, Abrikosov IA (2015) Computer simulations of glasses: the potential energy landscape. J Phys: Condens Matter 27:293201
170. Yin J, Wang Y, Chen JZY, Zhang P, Zhang L (2020) Construction of a pathway map on a complicated energy landscape. Phys Rev Lett 124:090601
171. Shires BWB, Pickard CJ (2021) Visualizing energy landscapes through manifold learning. Phys Rev X 11:041026
172. Angell CA (1995) Formation of glasses from liquids and biopolymers. Science 267:1924
173. Sastry S, Debenedetti PG, Stillinger FH (1998) Signatures of distinct dynamical regimes in the energy landscape of a glass-forming liquid. Nature 393:554

174. Sastry S (2001) The relationship between fragility, configurational entropy and the potential energy landscape of glass-forming liquids. Nature 409:164
175. Debenedetti PG, Stillinger FH (2001) Supercooled liquids and the glass transition. Nature 410:259
176. Garstecki P, Hoang TX, Cieplak M (1999) Energy landscapes, supergraphs, and "folding funnels" in spin systems. Phys Rev E 60:3219
177. Zhou Q (2011) RandomWalk over basins of attraction to construct ising energy landscapes. Phys Rev Lett 106:180602
178. Samarakoon A, Sato TJ, Chen T, Chern G-W, Yang J, Klich I, Sinclair R, Zhou H, Lee S-H (2016) Aging, memory, and nonhierarchical energy landscape of spin jam. Proc Natl Acad Sci USA 113:11806
179. Jin Y, Yoshino H (2017) Exploring the complex free-energy landscape of the simplest glass by rheology. Nat Commun 8:14935
180. Malandro DL, Lacks DJ (1999) Relationships of shear-induced changes in the potential energy landscape to the mechanical properties of ductile glasses. J Chem Phys 110:4593
181. Lacks DJ, Osborne MJ (2004) Energy landscape picture of overaging and rejuvenation in a sheared glass. Phys Rev Lett 93:255501
182. Cao P, Short MP, Yip S (2019) Potential energy landscape activations governing plastic flows in glass rheology. Proc Natl Acad Sci USA 116:18790
183. Mortenson PN, Evans DA, Wales DJ (2002) Energy landscapes of model polyalanines. J Chem Phys 117:1363
184. Chen S, Ho M-H, Bullock RM, DuBois DL, Dupuis M, Rousseau R, Raugei S (2014) Computing free energy landscapes: application to Ni-based electrocatalysts with pendant amines for H_2 production and oxidation. ACS Catal 4:229
185. Frauenfelder H, Sligar SG, Wolynes PG (1991) The energy landscapes and motions of proteins. Science 254:1598
186. Wolynes PG, Onuchic JN, Thirumalai D (1995) Navigating the folding routes. Science 267:1619
187. Onuchic JN, Wolynes PG, Luthey-Schulten Z, Socci ND (1995) Toward an outline of the topography of a realistic protein-folding funnel. Proc Natl Acad Sci USA 92:3626
188. Henzler-Wildman K, Kern D (2007) Dynamic personalities of proteins. Nature 450:964
189. Dill KA, MacCallum JL (2012) The protein-folding problem, 50 years on. Science 338:1042
190. Kmiecik S, Gront D, Kolinski M, Wieteska L, Dawid AE, Kolinski A (2016) Coarse-grained protein models and their applications. Chem Rev 116:7898
191. Schiffrin B, Brockwell DJ, Radford SE (2017) Outer membrane protein folding from an energy landscape perspective. BMC Biol 15:123
192. Othayoth R, Thoms G, Li C (2020) An energy landscape approach to locomotor transitions in complex 3D terrain. Proc Natl Acad Sci USA 117:14987
193. Othayoth R, Xuan Q, Wang Y, Li C (2021) Locomotor transitions in the potential energy landscape-dominated regime. Proc R Soc B 288:20202734
194. Ballard AJ, Das R, Martiniani S, Mehta D, Sagun L, Stevenson JD, Wales DJ (2017) Energy landscapes for machine learning. Phys Chem Chem Phys 19:12585
195. Mehta P, Bukov M, Wang C-H, Day AGR, Richardson C, Fisher CK, Schwab DJ (2019) A high-bias, low-variance introduction to machine learning for physicists. Phys Rep 810:1
196. Turner JS, O'Halloran JA, Kalaidina E, Kim W, Schmitz AJ, Zhou JQ, Lei T, Thapa M, Chen RE, Case JB, Amanat F, Rauseo AM, Haile A, Xie X, Klebert MK, Suessen T, Middleton WD, Shi P-Y, Krammer F, Teefey SA, Diamond MS, Presti RM, Ellebedy AH (2021) SARS-CoV-2 mRNA vaccines induce persistent human germinal centre responses. Nature 596:109
197. Tartof SY, Slezak JM, Fischer H, Hong V, Ackerson BK, Ranasinghe ON, Frankland TB, Ogun OA, Zamparo JM, Gray S, Valluri SR, Pan K, Angulo FJ, Jodar L, McLaughlin JM (2021) Effectiveness of mRNA BNT162b2 COVID-19 vaccine up to 6 months in a large integrated health system in the USA: a retrospective cohort study. Lancet 398:1407

198. Chemaitelly H, Tang P, Hasan MR, AlMukdad S, Yassine HM, Benslimane FM, Al Khatib HA, Coyle P, Ayoub HH, Al Kanaani Z, Al Kuwari E, Jeremijenko A, Kaleeckal AH, Latif AN, Shaik RM, Abdul Rahim HF, Nasrallah GK, Al Kuwari MG, Al Romaihi HE, Butt AA, Al-Thani MH, Al Khal A, Bertollini R, Abu-Raddad LJ (2021) Waning of BNT162b2 vaccine protection against SARS-CoV-2 infection in qatar. N Engl J Med 385:e83
199. Goel RR, Painter MM, Apostolidis SA, Mathew D, Meng W, Rosenfeld AM, Lundgreen KA, Reynaldi A, Khoury DS, Pattekar A, Gouma S, Kuri-Cervantes L, Hicks P, Dysinger S, Hicks A, Sharma H, Herring S, Korte S, Baxter AE, Oldridge DA, Giles JR, Weirick ME, McAllister CM, Awofolaju M, Tanenbaum N, Drapeau EM, Dougherty J, Long S, D'Andrea K, Hamilton JT, McLaughlin M, Williams JC, Adamski S, Kuthuru O, The UPenn COVID Processing Unit, Frank I, Betts MR, Vella LA, Grifoni A, Weiskopf D, Sette A, Hensley SE, Davenport MP, Bates P, Prak ETL, Greenplate AR, Wherry EJ (2021) mRNA vaccines induce durable immune memory to SARS-CoV-2 and variants of concern. Science 374:1241
200. Gadala-Maria F, Acrivos A (1980) Shear-induced structure in a concentrated suspension of solid spheres. J Rheol 24:799
201. Fleming RM, Schneemeyer LF (1983) Transient electrical response of $K_{0.30}MoO_3$. Phys Rev B 28:6996
202. Sethna JP, Dahmen K, Kartha S, Krumhansl JA, Roberts BW, Shore JD (1993) Hysteresis and hierarchies: dynamics of disorder-driven first-order phase transformations. Phys Rev Lett 70:3347
203. Coppersmith SN, Jones TC, Kadanoff LP, Levine A, McCarten JP, Nagel SR, Venkataramani SC, Wu X (1997) Self-organized short-term memories. Phys Rev Lett 78:3983
204. Jonason K, Vincent E, Hammann J, Bouchaud JP, Nordblad P (1998) Memory and chaos effects in spin glasses. Phys Rev Lett 81:3243
205. Toiya M, Stambaugh J, Losert W (2004) Transient and oscillatory granular shear flow. Phys Rev Lett 93:088001
206. Keim NC, Nagel SR (2011) Generic transient memory formation in disordered systems with noise. Phys Rev Lett 107:010603
207. Fiocco D, Foffi G, Sastry S (2014) Encoding of memory in sheared amorphous solids. Phys Rev Lett 112:025702
208. Paulsen JD, Keim NC, Nagel SR (2014) Multiple transient memories in experiments on sheared non-Brownian suspensions. Phys Rev Lett 113:068301
209. Mungan M, Sastry S, Dahmen K, Regev I (2019) Networks and hierarchies: how amorphous materials learn to remember. Phys Rev Lett 123:178002
210. Charbonneau P, Morse PK (2021) Memory formation in jammed hard spheres. Phys Rev Lett 126:088001
211. Lindeman CW, Nagel SR (2021) Multiple memory formation in glassy landscapes. Sci Adv 7:eabg7133
212. Keim NC, Medina D (2022) Mechanical annealing and memories in a disordered solid. Sci Adv 8:eabo1614
213. Okuma S, Kawamura Y, Tsugawa Y (2012) Irreversibility and dislocation in periodically sheared vortex matter. J Phys Soc Jpn 81:114718
214. Dobroka M, Ienaga K, Kawamura Y, Kaneko S, Okuma S (2019) Competition between dynamic ordering and disordering for vortices driven by superimposed ac and dc forces. New J Phys 21:043007
215. Maegochi S, Dobroka M, Ienaga K, Kaneko S, Okuma S (2019) Time evolution of the vortex configuration associated with dynamic ordering detected by dc drive. J Phys: Conf Ser 1293:012023
216. Kibble T (2007) Phase-transition dynamics in the lab and the universe. Phys Today 60(9):47
217. del Campo A, Kibble TWB, Zurek WH (2013) Causality and non-equilibrium second-order phase transitions in inhomogeneous systems. J Phys: Condens Matter 25:404210
218. del Campo A, Zurek WH (2014) Universality of phase transition dynamics: topological defects from symmetry breaking. Int J Mod Phys A 29:1430018
219. Kibble TWB (1976) Topology of cosmic domains and strings. J Phys A: Math Gen 9:1387

220. Kibble TWB (1980) Some implications of a cosmological phase transition. Phys Rep 67:183
221. Zurek WH (1985) Cosmological experiments in superfluid helium? Nature 317:505
222. Zurek WH (1996) Cosmological experiments in condensed matter systems. Phys Rep 276:177
223. Hohenberg PC, Halperin BI (1977) Theory of dynamic critical phenomena. Rev Mod Phys 49:435
224. Tinkham M (2004) Introduction to superconductivity, 2nd edn. Dover Publications, New York
225. Bennemann KH, Ketterson JB (eds) (2008) Superconductivity: vol. 1: conventional and unconventional superconductors. Springer, Berlin
226. van Delft D, Kes P (2010) The discovery of superconductivity. Phys Today 63(9):38
227. Abrikosov AA (1957) On the magnetic properties of superconductors of the second group. Sov Phys JETP 5:1174
228. Blatter G, Feigel'man MV, Geshkenbein VB, Larkin AI, Vinokur VM (1994) Vortices in high-temperature superconductors. Rev Mod Phys 66:1125
229. Brandt EH (1995) The flux-line lattice in superconductors. Rep Prog Phys 58:1465
230. Bending SJ (1999) Local magnetic probes of superconductors. Adv Phys 48:449
231. Essmann U, Träuble H (1967) The direct observation of individual flux lines in type II superconductors. Phys Lett A 24:526
232. Gammel PL, Bishop DJ, Dolan GJ, Kwo JR, Murray CA, Schneemeyer LF, Waszczak JV (1987) Observation of hexagonally correlated flux quanta in $YBa_2Cu_3O_7$. Phys Rev Lett 59:2592
233. Bolle CA, Gammel PL, Grier DG, Murray CA, Bishop DJ, Mitzi DB, Kapitulnik A (1991) Observation of a commensurate array of flux chains in tilted flux lattices in Bi-Sr-Ca-Cu-O single crystals. Phys Rev Lett 66:112
234. Dai H, Yoon S, Liu J, Budhani RC, Lieber CM (1994) Simultaneous observation of columnar defects and magnetic flux lines in high-temperature $Bi_2Sr_2CaCu_2O_8$ superconductors. Science 265:1552
235. Fasano Y, Menghini M (2008) Magnetic-decoration imaging of structural transitions induced in vortex matter. Supercond Sci Technol 21:023001
236. Hess HF, Robinson RB, Dynes RC, Valles JM Jr, Waszczak JV (1989) Scanning-tunneling-microscope observation of the abrikosov flux lattice and the density of states near and inside a fluxoid. Phys Rev Lett 62:214
237. Fischer Ø, Kugler M, Maggio-Aprile I, Berthod C, Renner C (2007) Scanning tunneling spectroscopy of high-temperature superconductors. Rev Mod Phys 79:353
238. Guillamón I, Córdoba R, Sesé J, De Teresa JM, Ibarra MR, Vieira S, Suderow H (2014) Enhancement of long-range correlations in a 2D vortex lattice by an incommensurate 1D disorder potential. Nat Phys 10:851
239. Harada K, Matsuda T, Bonevich J, Igarashi M, Kondo S, Pozzi G, Kawabe U, Tonomura A (1992) Real-time observation of vortex lattices in a superconductor by electron microscopy. Nature 360:51
240. Matsuda T, Harada K, Kasai H, Kamimura O, Tonomura A (1996) Observation of dynamic interaction of vortices with pinning centers by Lorentz microscopy. Science 271:1393
241. Harada K, Kamimura O, Kasai H, Matsuda T, Tonomura A, Moshchalkov VV (1996) Direct observation of vortex dynamics in superconducting films with regular arrays of defects. Science 274:1167
242. Matsuda T, Kamimura O, Kasai H, Harada K, Yoshida T, Akashi T, Tonomura A, Nakayama Y, Shimoyama J, Kishio K, Hanaguri T, Kitazawa K (2001) Oscillating rows of vortices in superconductors. Science 294:2136
243. Vu LN, Wistrom MS, Van Harlingen DJ (1993) Imaging of magnetic vortices in superconducting networks and clusters by scanning SQUID microscopy. Appl Phys Lett 63:1693
244. Kirtley JR, Ketchen MB, Stawiasz KG, Sun JZ, Gallagher WJ, Blanton SH, Wind SJ (1995) High-resolution scanning SQUID microscope. Appl Phys Lett 66:1138
245. Embon L, Anahory Y, Jelić ŽL, Lachman EO, Myasoedov Y, Huber ME, Mikitik GP, Silhanek AV, Milošević MV, Gurevich A, Zeldov E (2017) Imaging of super-fast dynamics and flow instabilities of superconducting vortices. Nat Commun 8:85

246. Moser A, Hug HJ, Parashikov I, Stiefel B, Fritz O, Thomas H, Baratoff A, Güntherodt H-J, Chaudhari P (1995) Observation of single vortices condensed into a vortex-glass phase by magnetic force microscopy. Phys Rev Lett 74:1847
247. Roseman M, Grütter P (2001) Estimating the magnetic penetration depth using constant-height magnetic force microscopy images of vortices. New J Phys 3:24
248. Oral A, Bending SJ, Henini M (1996) Real-time scanning Hall probe microscopy. Appl Phys Lett 69:1324
249. Oral A, Barnard JC, Bending SJ, Kaya II, Ooi S, Tamegai T, Henini M (1998) Direct observation of melting of the vortex solid in $Bi_2Sr_2CaCu_2O_{8+\delta}$ single crystals. Phys Rev Lett 80:3610
250. Grigorenko A, Bending S, Tamegai T, Ooi S, Henini M (2001) A one-dimensional chain state of vortex matter. Nature 414:728
251. Field SB, James SS, Barentine J, Metlushko V, Crabtree G, Shtrikman H, Ilic B, Brueck SRJ (2002) Vortex configurations, matching, and domain structure in large arrays of artificial pinning centers. Phys Rev Lett 88:067003
252. Goa PE, Hauglin H, Baziljevich M, Il'yashenko E, Gammel PL, Johansen TH (2001) Real-time magneto-optical imaging of vortices in superconducting $NbSe_2$. Supercond Sci Technol 14:729
253. van der Beek CJ, Colson S, Indenbom MV, Konczykowski M (2000) Supercooling of the disordered vortex lattice in $Bi_2Sr_2CaCu_2O_{8+\delta}$. Phys Rev Lett 84:4196
254. Paltiel Y, Zeldov E, Myasoedov Y, Rappaport ML, Jung G, Bhattacharya S, Higgins MJ, Xiao ZL, Andrei EY, Gammel PL, Bishop DJ (2000) Instabilities and disorder-driven first-order transition of the vortex lattice. Phys Rev Lett 85:3712
255. Fasano Y, Menghini M, de la Cruz F, Paltiel Y, Myasoedov Y, Zeldov E, Higgins MJ, Bhattacharya S (2002) Order-disorder phase transition in $NbSe_2$: absence of amorphous vortex matter. Phys Rev B 66:020512(R)
256. Okuma S, Kashiro K, Suzuki Y, Kokubo N (2008) Order-disorder transition of vortex matter in a-Mo_xGe_{1-x} films probed by noise. Phys Rev B 77:212505
257. Bermúdez MM, Eskildsen MR, Bartkowiak M, Nagy G, Bekeris V, Pasquini G (2015) Dynamic reorganization of vortex matter into partially disordered lattices. Phys Rev Lett 115:067001
258. Gammel PL, Schneemeyer LF, Waszczak JV, Bishop DJ (1988) Evidence from mechanical measurements for flux-lattice melting in single-crystal $YBa_2Cu_3O_7$ and $Bi_{2.2}Sr_2Ca_{0.8}Cu_2O_8$. Phys Rev Lett 61:1666
259. Cubitt R, Forgan EM, Yang G, Lee SL, Paul DMcK, Mook HA, Yethiraj M, Kes PH, Li TW, Menovsky AA, Tarnawski Z, Mortensen K (1993) Direct observation of magnetic flux lattice melting and decomposition in the high-T_c superconductor $Bi_{2.15}Sr_{1.95}CaCu_2O_{8+x}$. Nature 365:407
260. Zeldov E, Majer D, Konczykowski M, Geshkenbein VB, Vinokur VM, Shtrikman H (1995) Thermodynamic observation of first-order vortex-lattice melting transition in $Bi_2Sr_2CaCu_2O_8$. Nature 375:373
261. Soibel A, Zeldov E, Rappaport M, Myasoedov Y, Tamegai T, Ooi S, Konczykowski M, Geshkenbein VB (2000) Imaging the vortex-lattice melting process in the presence of disorder. Nature 406:282
262. Guillamón I, Suderow H, Fernández-Pacheco A, Sesé J, Córdoba R, De Teresa JM, Ibarra MR, Vieira S (2009) Direct observation of melting in a two-dimensional superconducting vortex lattice. Nat Phys 5:651
263. Roy I, Dutta S, Choudhury ANR, Basistha S, Maccari I, Mandal S, Jesudasan J, Bagwe V, Castellani C, Benfatto L, Raychaudhuri P (2019) Melting of the vortex lattice through intermediate hexatic fluid in an a-MoGe thin film. Phys Rev Lett 122:047001
264. Pippard AB (1969) A possible mechanism for the peak effect in type II superconductors. Philos Mag 19:217
265. Kes PH, Tsuei CC (1981) Collective-flux-pinning phenomena in amorphous superconductors. Phys Rev Lett 47:1930

266. Wördenweber R, Kes PH, Tsuei CC (1986) Peak and history effects in two-dimensional collective flux pinning. Phys Rev B 33:3172
267. Higgins MJ, Bhattacharya S (1996) Varieties of dynamics in a disordered flux-line lattice. Physica C 257:232
268. Gammel PL, Yaron U, Ramirez AP, Bishop DJ, Chang AM, Ruel R, Pfeiffer LN, Bucher E, D'Anna G, Huse DA, Mortensen K, Eskildsen MR, Kes PH (1998) Structure and correlations of the flux line lattice in crystalline Nb through the peak effect. Phys Rev Lett 80:833
269. Henderson W, Andrei EY, Higgins MJ (1998) Plastic motion of a vortex lattice driven by alternating current. Phys Rev Lett 81:2352
270. Paltiel Y, Zeldov E, Myasoedov YN, Shtrikman H, Bhattacharya S, Higgins MJ, Xiao ZL, Andrei EY, Gammel PL, Bishop DJ (2000) Dynamic instabilities and memory effects in vortex matter. Nature 403:398
271. Marchevsky M, Higgins MJ, Bhattacharya S (2001) Two coexisting vortex phases in the peak effect regime in a superconductor. Nature 409:591
272. Troyanovski AM, van Hecke M, Saha N, Aarts J, Kes PH (2002) STM imaging of flux line arrangements in the peak effect regime. Phys Rev Lett 89:147006
273. Li G, Andrei EY, Xiao ZL, Shuk P, Greenblatt M (2006) Onset of motion and dynamic reordering of a vortex lattice. Phys Rev Lett 96:017009
274. Iavarone M, Di Capua R, Karapetrov G, Koshelev AE, Rosenmann D, Claus H, Malliakas CD, Kanatzidis MG, Nishizaki T, Kobayashi N (2008) Effect of magnetic impurities on the vortex lattice properties in $NbSe_2$ single crystals. Phys Rev B 78:174518
275. Rumi G, Sánchez JA, Elías F, Maldonado RC, Puig J, Bolecek NRC, Nieva G, Konczykowski M, Fasano Y, Kolton AB (2019) Hyperuniform vortex patterns at the surface of type-II superconductors. Phys Rev Res 1:033057
276. Llorens JB, Guillamón I, Serrano IG, Córdoba R, Sesé J, De Teresa JM, Ibarra MR, Vieira S, Ortuño M, Suderow H (2020) Disordered hyperuniformity in superconducting vortex lattices. Phys Rev Res 2:033133
277. Puig J, Elías F, Sánchez JA, Maldonado RC, Rumi G, Nieva G, Pedrazzini P, Kolton AB, Fasano Y (2022) Anisotropic suppression of hyperuniformity of elastic systems in media with planar disorder. Commun Mater 3:32
278. Torquato S, Stillinger FH (2003) Local density fluctuations, hyperuniformity, and order metrics. Phys Rev E 68:041113
279. Torquato S (2018) Hyperuniform states of matter. Phys Rep 745:1
280. Gabrielli A, Joyce M, Labini FS (2002) Glass-like universe: real-space correlation properties of standard cosmological models. Phys Rev D 65:083523
281. Jiao Y, Lau T, Hatzikirou H, Meyer-Hermann M, Corbo JC, Torquato S (2014) Avian photoreceptor patterns represent a disordered hyperuniform solution to a multiscale packing problem. Phys Rev E 89:022721
282. Hexner D, Levine D (2015) Hyperuniformity of critical absorbing states. Phys Rev Lett 114:110602
283. Tjhung E, Berthier L (2015) Hyperuniform density fluctuations and diverging dynamic correlations in periodically driven colloidal suspensions. Phys Rev Lett 114:148301
284. Weijs JH, Jeanneret R, Dreyfus R, Bartolo D (2015) Emergent hyperuniformity in periodically driven emulsions. Phys Rev Lett 115:108301
285. Tjhung E, Berthier L (2016) Criticality and correlated dynamics at the irreversibility transition in periodically driven colloidal suspensions. J Stat Mech 033501
286. Hexner D, Levine D (2017) Noise, diffusion, and hyperuniformity. Phys Rev Lett 118:020601
287. Weijs JH, Bartolo D (2017) Mixing by unstirring: hyperuniform dispersion of interacting particles upon chaotic advection. Phys Rev Lett 119:048002
288. Wilken S, Guerra RE, Pine DJ, Chaikin PM (2020) Hyperuniform structures formed by shearing colloidal suspensions. Phys Rev Lett 125:148001

Chapter 2
Experimental

2.1 Sample Preparation

In this thesis, we used two types of amorphous (a-)Mo_xGe_{1-x} film with a strip shape and a cross shape. Each sample was fabricated by rf magnetron sputtering deposition onto a Si substrate. Figure 2.1a shows schematic illustration of the setup of the vacuum chamber for the sputtering. To attach the 8×8 mm^2 Si substrate to the sample holder shown in Fig. 2.1b, we used Ag paste that provides good thermal contact. The substrate was covered by a Cu mask to design the shape of the film. To estimate the sample thickness, a quartz oscillator, whose oscillating frequency changes depending on deposition amount, was also placed close to the sample. Then, we evacuated the vacuum chamber to a pressure of 10^{-7} Torr by using a rotary pump and a turbo molecular pump (TMP). To obtain ultra-high vacuum, the chamber was baked for 12 h at 80 °C. After that, the sputtering was performed in an Ar atmosphere. The Ar ions were accelerated by an electric field applied between the sample and MoGe target and they sputtered the MoGe particles. Then, the MoGe film was deposited on the Si substrate. During the sputtering, the sample holder was rotated at 240 rpm to obtain the uniform sample. In addition, coolant water was flowing through the sample holder and target holder shown in Fig. 2.1c to keep the temperature below 10 °C, which avoids the crystallization of Mo. Figure 2.1d shows the typical diffraction pattern of the thus prepared sample obtained by transmission electron microscopy. The halo pattern without specific peaks indicates that the sample has no long-range order and the amorphous sample is obtained.

2.2 Sample Characteristics

We prepared the strip-shaped sample and the cross-shaped sample. The sample geometries with nominal size are shown in Fig. 2.2a, b. Table 2.1 summarizes the characteristics of the samples such as Mo concentration x, film thickness,

S. Maegochi, *Nonequilibrium Phase Transitions in Driven Vortex Matter*, Springer Theses,
https://doi.org/10.1007/978-981-97-2987-6_2

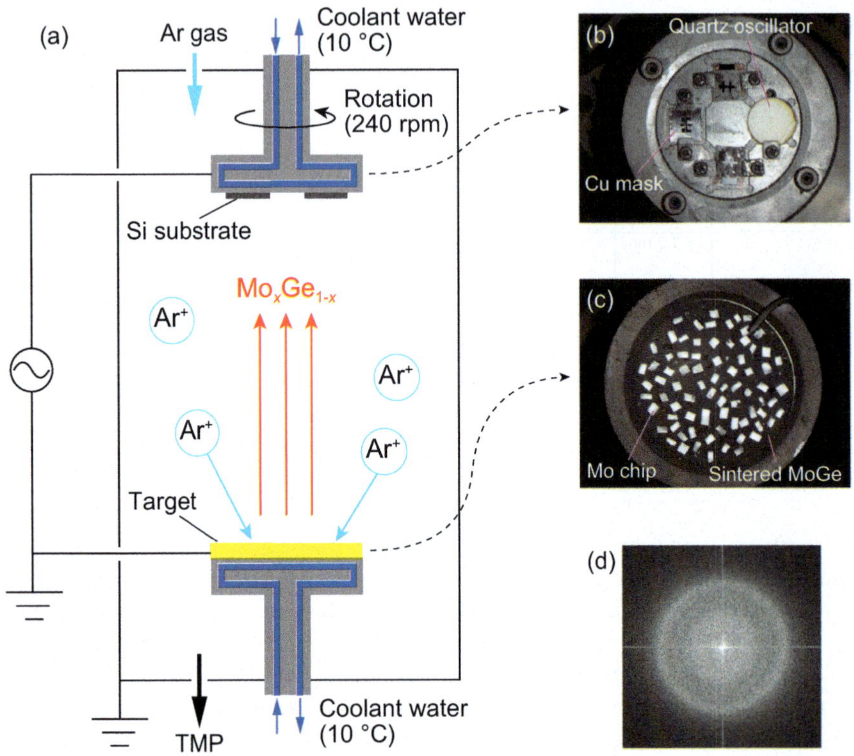

Fig. 2.1 The setup for sample preparation. **a** The schematic illustration of the vacuum chamber. Ar ions accelerated by the bias sputters the MoGe particles. Coolant water is applied to keep the temperature low, which avoids the crystallization of Mo. **b** Image of the sample holder. The Si substrate and quartz oscillator are placed on the sample holder. The substrate is covered by the Cu mask to obtain the desired shape of the film. During the sputtering, the sample holder is rotated at 240 rpm in order to obtain the uniform film. **c** Image of the target holder. The sintered $Mo_{0.7}Ge_{0.3}$ and Mo chips are placed on the target holder. **d** The diffraction pattern of the sample obtained by transmission electron microscopy. The halo pattern without specific peaks indicates that the amorphous sample is obtained

normal-state resistivity ρ_n, superconducting transition temperatures T_{c0} and T_c, and the peak field B_p.

We measured linear resistivity ρ under zero magnetic field ($B = 0$) using a standard four-probe method. Figure 2.3a, c show the temperature dependence of ρ for the strip-shaped film and the cross-shaped film, respectively, with measuring current of 10 μA. The resistivity in the normal state at 7 K is $\rho_n = 2.10$ μΩm for the strip-shaped film and $\rho_n = 1.60$ and 1.67 μΩm in x and y directions, respectively, for the cross-shaped film. The difference of ρ_n between x and y directions may result from the difference of actual sample size from the nominal size. The superconducting transition temperatures T_{c0} and T_c were determined as the temperature at which $\rho(T)$ falls to $0.95\rho_n$ and to the measurement limit of $V = 10^{-8}$ V, respectively.

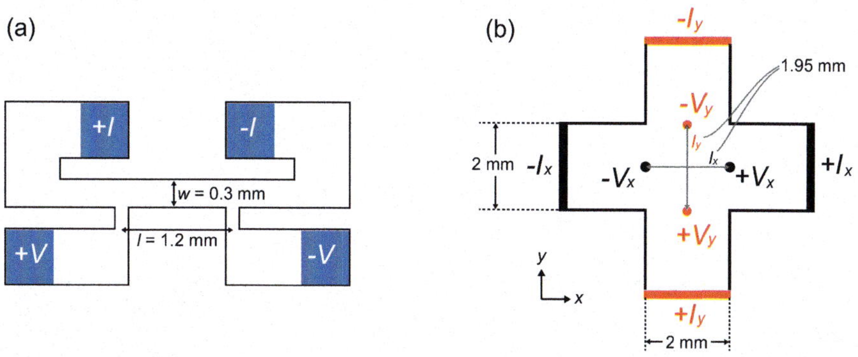

Fig. 2.2 The geometries of samples. **a** The strip-shaped sample. **b** The cross-shaped sample. Nominal size is shown in the figures

Table 2.1 Characteristics of the a-Mo_xGe_{1-x} samples

	Stripe-shaped film	Cross-shaped film	
		x direction	y direction
x	0.78	0.78	
Thickness [nm]	330	280	
ρ_n [μΩm]	2.10	1.60	1.67
T_{c0} [K]	6.4	6.6	6.6
T_c [K]	6.3	6.2	6.2
B_p [T]	4.0 (at 4.1 K)	2.2 (at 3.6 K)	2.2 (at 3.6 K)

The superconducting transition width of the cross-shaped film is slightly wider than that of the strip-shaped film possibly because the large sample size of the cross-shaped film leads to a little inhomogeneity. However, we consider that this possible inhomogeneity is negligible under the magnetic field since the dynamics of vortices discussed below, such as the depinning, are almost the same in both directions of the film.

For both samples, the vortex system was prepared by applying the field B perpendicular to the plane of the film. Figure 2.3b, d show the depinning current density $J_d(B)$ for the strip-shaped sample at 4.1 K and the cross-shaped sample at 3.6 K, respectively, where the peak effect is observed. We obtain the peak field $B_p = 4.0$ T for the strip-shaped sample and $B_p = 2.2$ T in both x and y directions for the cross-shaped sample.

2.3 ^{4}He Cryostat

We used a ^{4}He cryostat to conduct the low-temperature experiments. Figure 2.4 schematically illustrates the structure of ^{4}He cryostat constituted by three glass Dewars. The outer glass Dewar is filled with liquid N_2 to avoid radiation heating.

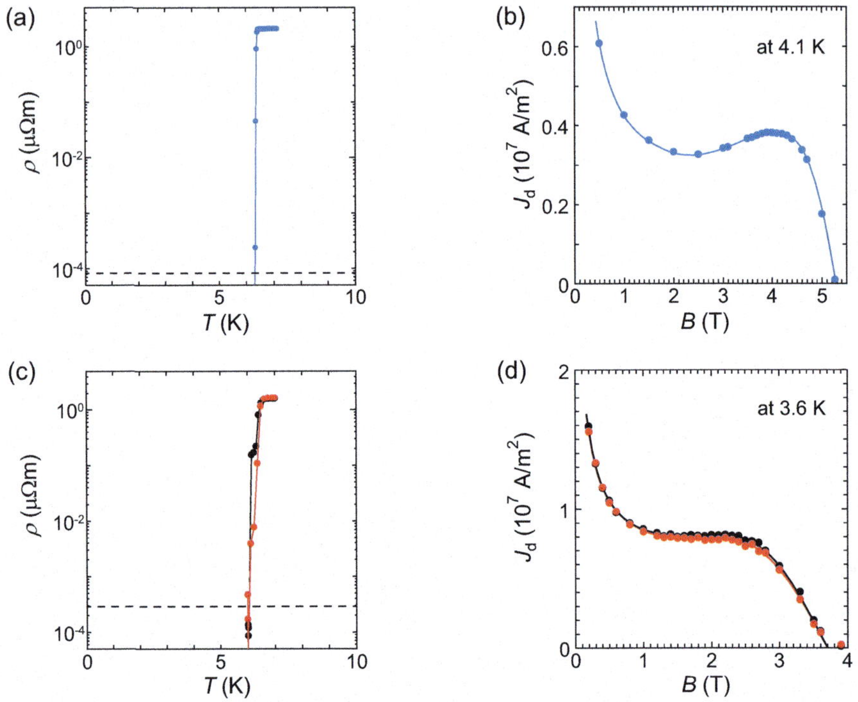

Fig. 2.3 Characterization of the a-Mo_xGe_{1-x} samples. **a**, **c** Resistivity $\rho(T)$ at zero magnetic field in a semi-log scale for the strip-shaped film (**a**) and the cross-shaped film (**b**). Horizontal dashed lines indicate the measurement limit of 10^{-8} V with measuring current of 10 μA. **b**, **d** The peak effect of the depinning current density $J_d(B)$ for the strip-shaped sample at 4.1 K (**b**) and the cross-shaped sample at 3.6 K (**d**). Black and red symbols in (**c**) and (**d**) represent the data measured in the x and y directions, respectively

The intermediate glass Dewar is filled with liquid ^{4}He. The superconducting magnet is placed in this reservoir and cooled down to a sufficient low temperature during the experiment. The sample, thermometer, and heater are set in the inner glass Dewar in which ^{4}He gas is liquefied. To adjust the temperature precisely, the pressure in these two Dewars filled with liquid ^{4}He are reduced to 690 Torr, corresponding to 4.1 K, for the strip-shaped film and 400 Torr, corresponding to 3.6 K, for the cross-shaped film. The sample is directly immersed into liquid ^{4}He to avoid the Joule heating.

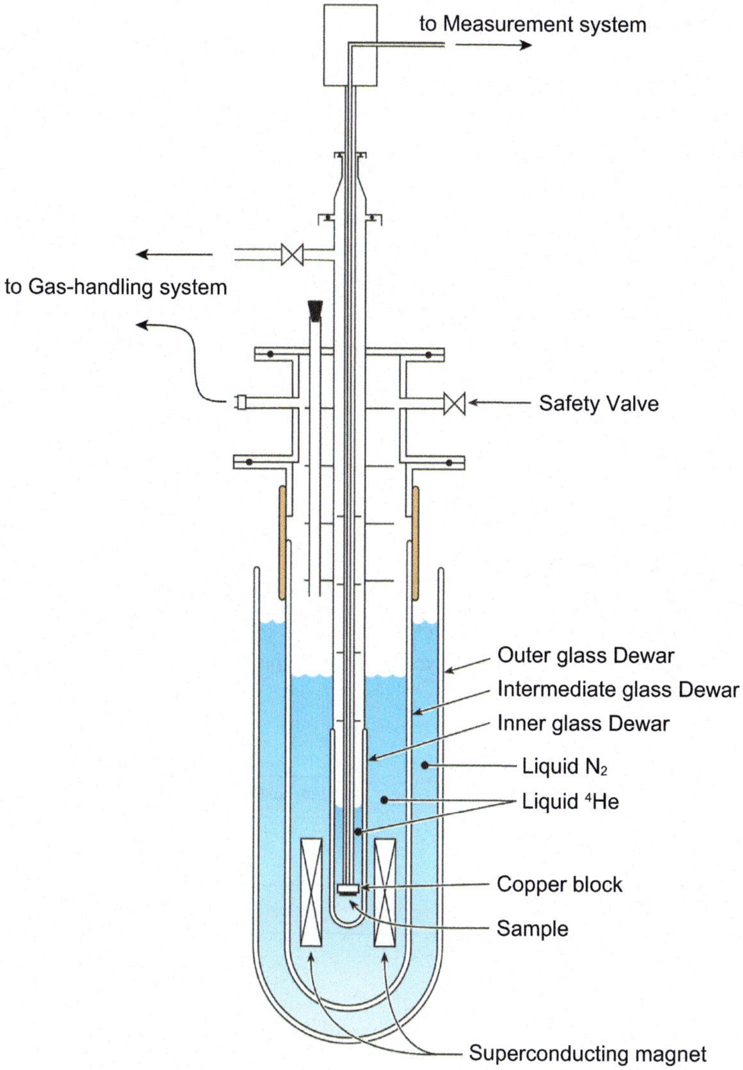

Fig. 2.4 Schematic illustration of the ^{4}He cryostat consisting of three glass Dewars

2.4 Measurement System

2.4.1 Measurements of the Steady-State Voltage

Dc current–voltage measurement

The dc I–V characteristics were measured by using a standard four-probe method. Figure 2.5 shows schematics of the circuit used in dc I–V measurement for the strip-shaped sample (a) and the cross-shaped sample (b). The vortices were driven by applying the dc current from a function generator (NF WF1965) and the voltage generated by the vortex motion was measured by a dc nanovoltmeter (Agilent 34420A). A standard resistor $R_{st} = 49.8124\ \Omega$ was connected in series with the sample. Hence, we can estimate the accurate value of the applied dc current from the voltage at the standard resistance R_{st}. To reduce the noise, we attached an attenuator (20 dB) to the function generator which attenuates the voltage to 1/10.

Mode-locking measurement

In Chap. 4, we carry out the mode-locking (ML) measurements. Details about ML are given in Chap. 4. Figure 2.6 shows schematics of the circuit used in the longitudinal

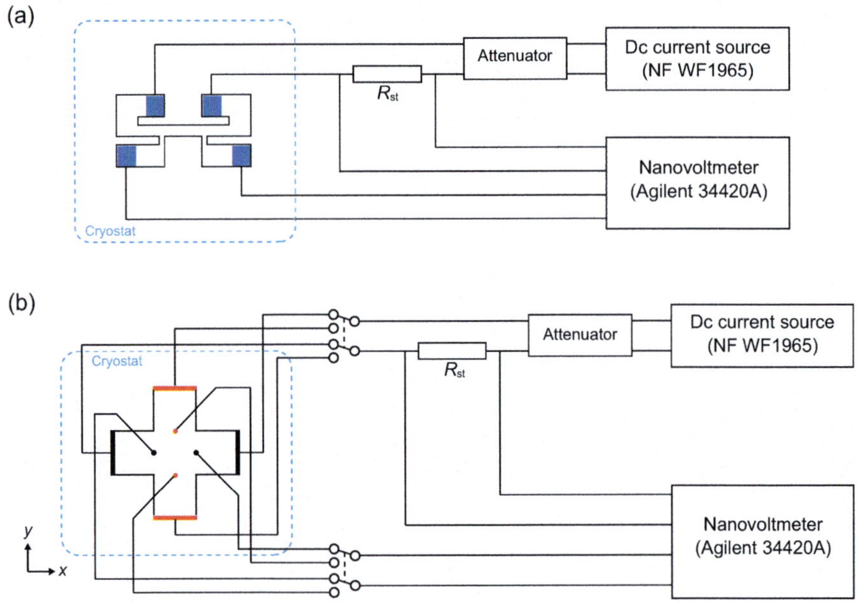

Fig. 2.5 The circuit for dc I–V measurement: For the strip-shaped sample (**a**) and the cross-shaped sample (**b**). The blue dashed line indicates the ^{4}He cryostat

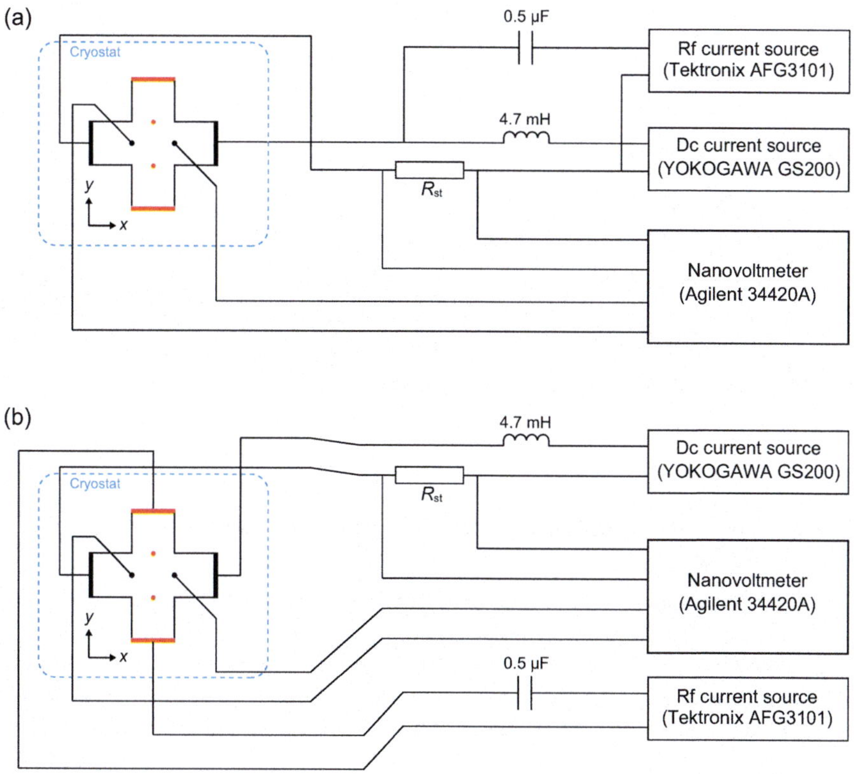

Fig. 2.6 The circuit for the longitudinal ML measurement (**a**) and the transverse ML measurement (**b**)

ML measurement (a) and the transverse ML measurement (b). The difference from the dc I–V measurement shown above is that the dc current source YOKOGAWA GS200 is used instead of NF WF1965 and rf current source (Tektronix AFG3101) is added in order to superimpose the rf current on the dc current. In the longitudinal ML measurement, the rf current was applied in the direction parallel to the dc current, while in the transverse ML measurement, the rf current was applied in the direction perpendicular to the dc current. To prevent rf current from entering the dc current source and vice versa, we inserted a capacitor of 0.5 μF and an inductor of 4.7 mH in front of rf and dc current source, respectively.

Transverse current–voltage measurement

In Chap. 5, we measure the transverse I–V characteristics. Figure 2.7 shows the schematic illustration of the circuit used in the transverse I–V measurement. The dc current source (YOKOGAWA GS200) is added to the circuit for the dc I–V measurement shown above.

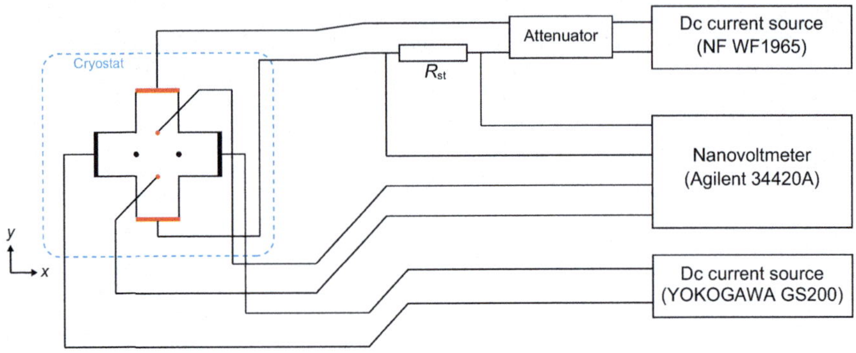

Fig. 2.7 The circuit for transverse I–V measurement

2.4.2 *Measurements of the Time-Dependent Voltage*

Figure 2.8 shows the circuit for the time-dependent voltage $V(t)$ measurements used in Chaps. 3, 6 (a) and Chap. 4 (b). The function generator WF1965 (NF) or AFG3251 (Tektronix) was used to drive vortices. To measure $V(t)$, we used an oscilloscope (Rohde and Schwarz RTO2024) whose time and voltage resolutions are about 100 ns and 10 μV, respectively. The voltage generated by the vortex motion was amplified

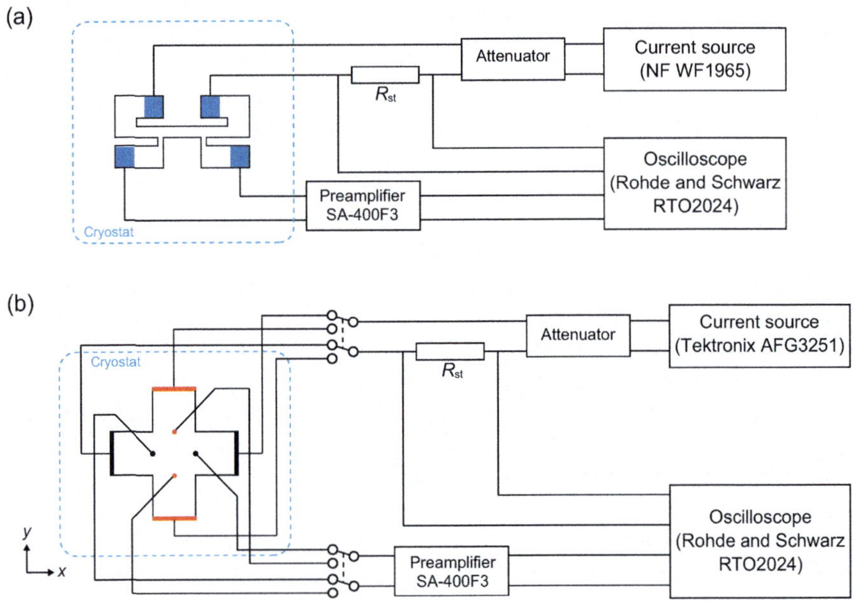

Fig. 2.8 The circuit for the time-dependent voltage $V(t)$ measurements used in Chaps. 3, 6 (**a**) and Chap. 4 (**b**)

by a factor of 100 by using a preamplifier (SA-400F3), which is enclosed in high permeability magnetic alloy (μ-metal) container to screen out the ambient radiation. The standard resistor $R_{st} = 49.8124\ \Omega$ was connected in series with the sample, so that we can estimate the accurate value of the applied dc current from the voltage at the standard resistance R_{st}. To reduce the noise, we also attached the attenuator (20 dB) to the function generator which attenuates the voltage to 1/10.

Chapter 3
Critical Behavior of RIT Driven by Particle Density as Well as Shear Amplitude

3.1 Introduction

As explained in Sect. 1.1.1, RIT is the nonequilibrium phase transition which is observed in systems with many interacting particles subjected to periodic shear [1]. Figure 3.1 shows a schematic phase diagram of RIT in a particle density n and shear amplitude d plane [2]. The diagram is drawn based on the results of experiments on colloidal suspensions [3, 4] and simulations [5–7], in which RIT is driven by d (shear-driven RIT). In other words, the boundary between reversible and irreversible phases shown by a solid line is traversed vertically as represented by an upward arrow. As seen in the diagram, RIT will be also observed by increasing the particle density n, and hence decreasing interparticle distance, at fixed d as shown by a right arrow. A natural question that comes to mind is whether the RIT driven by n (density-driven RIT) exhibits the same critical behavior and whether it belongs to the same universality class as the shear-driven RIT, namely, the DP universality class. Within the linear approximation that is usually valid in the vicinity of the transition, the critical behavior and the related critical exponent would not depend on the parameter that drives the transition. In fact, numerical studies using different computational models have found that the critical exponents are similar for the shear-driven RIT [8, 9] and the density-driven RIT [10, 11].

However, we consider that this is not trivial in practical systems. Several experiments investigating the critical phenomena of phase transitions [12, 13] have shown that the critical exponents of the transition take the different values depending on the control parameters. To our knowledge, there has been no experiments studying the critical behavior of the density-driven RIT. This is probably because in experimental systems where RIT has been intensively explored such as suspensions of colloids [3, 14] and emulsions [15, 16], it is difficult to carry out measurements with controllably changing the particle density n.

In this chapter, we explore the critical behavior of the density-driven RIT using the vortex system. The critical behavior of the shear-driven RIT has been already

S. Maegochi, *Nonequilibrium Phase Transitions in Driven Vortex Matter*, Springer Theses,
https://doi.org/10.1007/978-981-97-2987-6_3

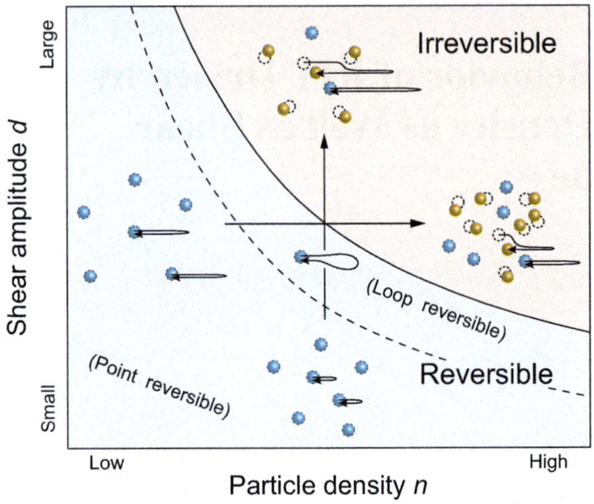

Fig. 3.1 Schematic phase diagram of RIT in the particle density n and shear amplitude d plane. The phase boundary of RIT is shown with the solid line. Blue and red shaded areas represent the reversible and irreversible states, respectively. The upward and right arrows denote the shear-driven and density-driven RIT, respectively. The dashed line separates the reversible regimes into the collisionless point-reversible state and the loop-reversible state with collisions between particles. Inset: Schematic illustration of the particle motion. Starting positions of particles are marked by dotted circles, while the particle positions at the end of one periodic drive are shown by closed circles. Blue and orange circles, respectively, depict the reversible and irreversible particles. Typical particle trajectories are shown schematically with arrows starting from the particles. Reprinted from Ref. [2] by S. Maegochi et al. licensed under CC BY 4.0

observed, which enables us to directly compare the critical behaviors between the shear-driven and density-driven RIT. The vortex density $n(\propto B)$ is proportional to the applied magnetic field B, so that we can well control n and the average intervortex spacing a_0 (Eq. 1.13).

For the shear-driven RIT [17], we have found that with decreasing d in the reversible phase ($d < d_c$), $\tau(d)$ drops significantly around a shear amplitude d_{c1} as shown with open blue squares in Fig. 1.20c. Since the value of d_{c1} is close to the average intervortex spacing a_0, combined with the recent computational work [5–7, 18], this implies the presence of a loop-reversible state with vortex-vortex collisions and a collisionless point-reversible state as drawn in Fig. 3.1. Therefore, we investigate the reversible regime in more detail.

3.2 Experiments

We used the strip-shaped a-Mo_xGe_{1-x} film presented in Sect. 2.2. All the experiments were conducted at 4.1 K. In this work, we used the value of B ranged from 2.0 T to 4.2 T in the peak-effect regime. The corresponding vortex density $(1/a_0)^2 (= \sqrt{3}B/2\Phi_0)$

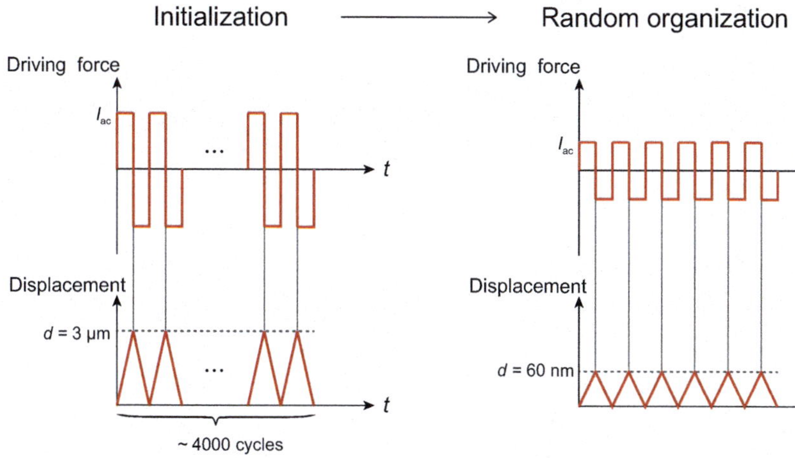

Fig. 3.2 Schematics of the experimental procedure consisting of initialization and random organization processes

is varied from $(1/35)^2$ nm^{-2} to $(1/24)^2$ nm^{-2}, where $a_0 = (2\Phi_0/\sqrt{3}B)^{1/2}$ is the average intervortex spacing.

Figure 3.2 shows the experimental procedure. At each B, we first initialized the configuration of vortices by applying the square current I_{ac} and shearing them for more than 4000 cycles with shear amplitude of $d = 3$ μm, which is sufficiently large compared with the critical shear amplitude $d_c \approx 45$ nm of the shear-driven RIT obtained in our previous work [17]. For initialization, the frequency f of I_{ac} was fixed at 4 kHz and the amplitude of I_{ac} was adjusted to yield the steady-state voltage V^∞ that gives $d = 3$ μm. Through this process, we obtained the random initial vortex configuration by a use of the diffusivity in the irreversible state. Then, to observe the random organization, we applied I_{ac} to the initial state and measured the time-dependent voltage $V(t)$ using the oscilloscope. We adjusted the amplitude of I_{ac} to yield a desired value of V^∞ that gives $d = 60$ nm, where the value of f was fixed at 200 kHz. According to the equation $d = v^\infty/2f$ (Eq. 1.18), a use of constant d with fixed f leads to the same vortex velocity v^∞ for different B. This enables us to ignore the effect of the velocity on RIT [19].

3.3 Results and Discussion

3.3.1 The Density-Driven RIT

Figure 3.3a, b show the representative voltage responses $|V(t)|/V^\infty$, which is normalized by steady-state voltage $V^\infty[\equiv V(t \to \infty)]$, to the periodic shear with $d = 60$ nm for $B = 2.0$, 3.0, 3.2 T and $B = 4.2$, 3.8, 3.4 T from top to bottom,

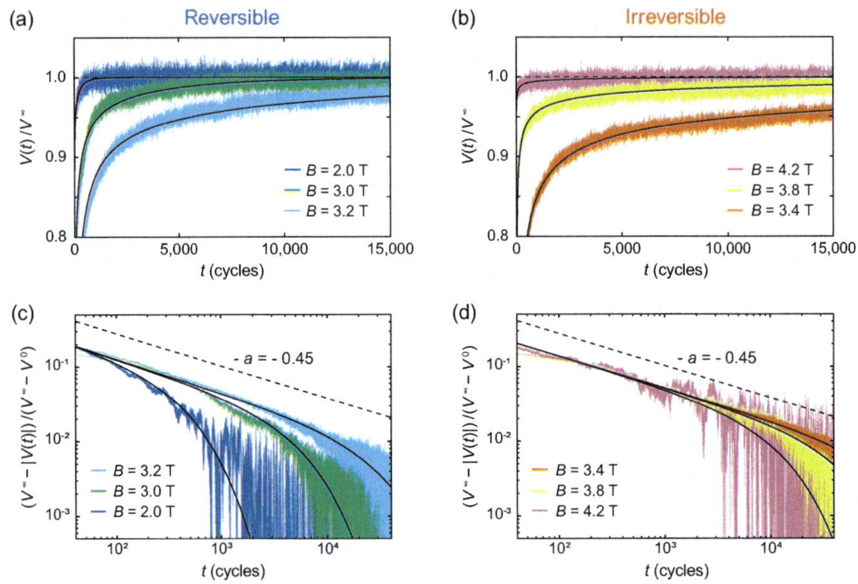

Fig. 3.3 Random organization of periodically sheared vortices around the density-driven RIT. **a, b** Time evolution of voltages $|V(t)|/V^{\infty}$ induced by vortices subjected to periodic shear with fixed amplitude of $d = 60$ nm at various fields: **a** $B = 2.0$, 3.0, and 3.2 T, and **b** $B = 4.2$, 3.8, and 3.4 T from top to bottom. For clarity, only the amplitude of the voltage pulse, $|V(t)|$, is plotted. Horizontal dashed lines mark the steady-state value of unity. **c, d** Replots of the data shown in (**a**, **b**) as $(V^{\infty} - |V(t)|)/(V^{\infty} - V^{0})$ versus t in a double logarithmic scale. **c** $B = 2.0$, 3.0, and 3.2 T and **d** $B = 4.2$, 3.8, and 3.4 T from bottom to top. Dashed lines in (**c, d**) show the power law of t with the exponent $-a = -0.45$. Solid lines in (**a**–**d**) represent the results of the fit to Eq. 3.1. Reprinted from Ref. [2] by S. Maegochi et al. licensed under CC BY 4.0

respectively. For clarity, vertical lines of each ac voltage pulse are taken away from the plots. In addition, the large voltage region, $|V(t)|/V^{\infty} > 0.8$, is plotted to easily distinguish the voltage curves.

All data of $|V(t)|/V^{\infty}$ show a common trend of monotonic increase with the number of shear cycles t and relaxation to a steady-state value, $|V(t)|/V^{\infty} = 1$, marked by a horizontal dashed line. The increasing relaxation of the voltage, namely, the average velocity of vortices is explained from the random organization of vortices as follows: A small initial voltage indicates that the flow of vortices at $t \sim 0$ is a pinning-dominated disordered flow with many vortex-vortex collisions, where the vortices cannot move easily. With increasing the number of shear cycles, $V(t)$ increases monotonically because the collided vortices change their distribution to avoid next collisions and rearrange into more organized configuration. Thus, less collisions occur in the steady state and vortices become more mobile than in the initial state.

The $|V(t)|/V^{\infty}$ curves in Fig. 3.3a show the longer relaxation for higher B, while those in Fig. 3.3b show the longer relaxation for lower B, indicating the presence of a peak of the relaxation time $\tau(B)$ around $B = 3.2$–3.4 T. To see the critical behavior of RIT, we extract τ from fit of the transient voltage $|V(t)|$ to the following relaxation function presented in Refs. [8, 9, 20]:

$$|V(t)| = V^{\infty} - (V^{\infty} - V^{0})\frac{e^{-t/\tau}}{t^{\alpha}}. \tag{3.1}$$

Here, V^0 and V^{∞} are the initial and steady-state voltage amplitudes, respectively, and τ is the relaxation time at which a power-law decay with an exponent α crosses over to an exponential relaxation. A power of t in the denominator is only relevant for $\tau/t \gg 1$, that is, close to the transition with diverging τ or at $t \sim 0$. This is indeed confirmed in Fig. 3.3c, d, where the data shown in Fig. 3.3a, b are replotted in the form of $(V^{\infty} - |V(t)|)/(V^{\infty} - V^{0})$, which should behave as $e^{-\tau/t}/t^{\alpha}$, versus t in a double logarithmic scale. For $B = 3.2$ and 3.4 T, the exponential decay is negligible due to sufficiently long τ and the power-law decay is clearly observed with a slope of $-\alpha = -0.45$ shown by dashed lines. The obtained value of $\alpha = 0.45 \pm 0.05$ is close to the critical exponent $\alpha = 0.40 \pm 0.05$ obtained for the shear-driven RIT [17] as well as the theoretical value $\alpha \approx 0.45$ for the 2D DP universality class [21], which predicts a power-law decay of the fraction of active (irreversible) particles as Eq. 1.2. The results of the fit to Eq. 3.1 using $\alpha = 0.45$ are shown with solid lines in Fig. 3.3a–d.

Figure 3.4 shows $\tau(B)$ determined from the fits, where closed and open circles are $\tau(B)$ for $B \leq 3.2$ T and $B \geq 3.4$ T, respectively. We find that τ diverges at $B_c = 3.26$ T as marked by a vertical dashed line, indicative of the density-driven RIT. The mean intervortex distance at B_c is $a_{0c} \approx 27$ nm. The inset to Fig. 3.4 shows the log-log plot of τ versus $|B - B_c|$, where symbols correspond to those in the main panel. The solid lines in the main panel and the inset display the results of the fit to the power law, $\tau \propto |B - B_c|^{-\nu}$, where we obtain the best fits for $\nu = 1.32 \pm 0.10$. This value is, within errors, again in agreement with the value obtained for the shear-driven RIT, $\nu = 1.38 \pm 0.08$ [17], and the theoretical one, $\nu = 1.295 \pm 0.006$, for the DP universality class in 2D [21]. The results demonstrate that RIT exhibits the same critical behavior irrespective of the control parameters that drive the transition, B and d, and that both types of RIT belong to the same universality class as the absorbing phase transition in 2D DP.

To our knowledge, this work is the first experimental observation of the density-driven RIT with critical behaviors. Meanwhile, the shear-driven RIT has been observed in several experimental systems. In order to compare the critical behaviors of τ in different experimental systems, we plot τ in Fig. 3.5 as a function of the dimensionless control parameters around the shear-driven RIT observed in colloidal suspensions [8], soft glasses [22], and dislocations in small crystals [23], together with our data around the density-driven RIT and the shear-driven RIT. The quantities, $|B - B_c|/B_c$, $|d - d_c|/d_c$, $|\gamma - \gamma_c|/\gamma_c$, and $|\sigma - \sigma_c|/\sigma_c$, on the horizontal axis are the normalized distance from the critical point, where γ represents the strain amplitude for colloidal suspensions and soft glasses, and σ stands for the stress for dislocations. The data shown with closed and open symbols are acquired in the

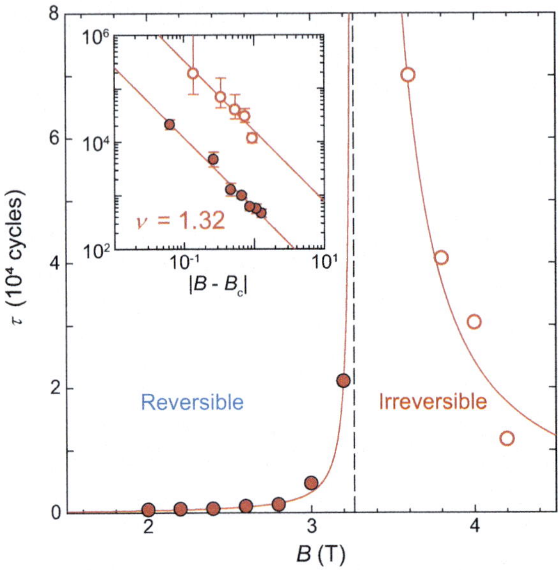

Fig. 3.4 The critical divergence of $\tau(B)$ around the density-driven RIT. Closed and open circles represent $\tau(B)$ for $B \leq 3.2$ T and $B \geq 3.4$ T, respectively. The vertical dashed line represents the critical point of RIT, $B_c = 3.26$ T, where the relaxation time diverges as the power law, $\tau \propto |B - B_c|^{-\nu}$, with $\nu = 1.32$ as shown by solid lines. Inset: τ versus $|B - B_c|$ in a double logarithmic scale. Symbols correspond to those in the main panel. Solid lines are the power-law fit whose slope gives the exponent $\nu = 1.32 \pm 0.10$. Reprinted from Ref. [2] by S. Maegochi et al. licensed under CC BY 4.0

reversible and irreversible states, respectively. Colored solid lines are the fits of τ to the power of each control parameter with exponent ν, whose value is displayed in the figure. Black solid and dashed lines have slopes of ν =1.295 and 1.110, respectively, which are expected for the 2D and 3D DP classes.

As mentioned before, our data for the density-driven and shear-driven RIT obey the power law marked by the red and blue solid lines, respectively, which have almost identical slopes of $\nu = 1.32 \pm 0.10$ and $\nu = 1.38 \pm 0.08$. These values are consistent with $\nu = 1.295 \pm 0.006$ in the 2D DP class, which agrees the fact that our vortex system is a 2D particlelike system. In contrast, the data for the colloidal suspensions and dislocations show $\nu = 1.1 \pm 0.3$ [8] and $\nu = 1.1$ [23], respectively. These values are slightly smaller than $\nu = 1.295$ for 2D DP, but close to $\nu = 1.110$ for 3D DP, which is also consistent with the fact that the colloidal suspensions and dislocations are 3D particlelike systems. Meanwhile, simulation of periodically driven dislocations in the 2D system has reported $\nu = 1.375$ [24] close to the value of 2D DP, where the dislocations always self-organize into reversible states both sides of the transition. The result of the soft glass is an exception because $\nu = 1.1 \pm 0.3$ [22] is close to the value of 3D DP although the soft glass is the 2D system. The authors of Ref. [22] argue that RIT they observed in the soft glass may fall into another universality class known as the conserved DP class [25–27], which predicts $\nu = 1.225 \pm 0.029$ in 2D.

Figure 3.5 also shows that the extent of the critical region for all the systems is ranged up to at least 0.3, which is much wider than that of ordinary critical phenomena in equilibrium [28, 29]. The critical exponents ν and α in this work do not depend on the control parameters B and d. This is somewhat surprising because the critical region observed here is too large to apply the linear approximation around the transition.

3.3.2 *Two Types of Reversible States*

Next, we look the details on the data of $\tau(d)$ in the reversible regime of the shear-driven RIT, which are shown with closed blue squares in Fig. 3.5. In Ref. [17], we have found that, when we decrease d in the reversible regime down to around 25 nm($\equiv d_{c1}$) close to the intervortex distance $a_0 \approx 26$ nm at 3.5 T, we observe the significant reduction and downward deviation of $\tau(d)$ from the power law, $\tau \propto (d_c - d)^{-\nu}$, which is highlighted by a dotted line in the main panel of Fig. 3.5. Here, we add the new data obtained for $d < d_{c1}$ and analyze the behavior of $\tau(d)$ below d_c in detail. The inset of Fig. 3.5 shows an enlarged view of the square area in the main panel. In the plot, we note the trivial fact that the dimensionless distance from the critical point in the reversible regime, $|d - d_c|/d_c$ never exceeds 1 since that corresponds to the limit of $d = 0$. After the sharp downward deviation at d_{c1}, $\tau(d)$ for $d < d_{c1}$ again exhibits the power-law divergence at the common critical point $d_c(= 45.2$ nm) as shown by the blue solid line. The power-low exponent $\nu = 1.30 \pm 0.18$ for $d < d_{c1}$ is, within errors, close to $\nu = 1.38 \pm 0.08$ obtained at $d_{c1} < d < d_c$ and $\nu = 1.295 \pm 0.006$ for 2D DP. This result implies the presence of two distinct reversible states separated at d_{c1} and indicates that they show identical critical behaviors around the common critical point d_c with the same critical exponent ν.

Recently, two distinct reversible regimes have been predicted by molecular dynamics simulations, where two regimes are termed the point-reversible and loop-reversible states [5, 6, 30]. In the point-reversible state, particles eventually organize into the arrangement in which they follow affine trajectories without collisions. When the affine displacement due to applied shear is subtracted from the trajectories, the remaining ones appear to be a point. This is why it is called the point-reversible state. On the other hand, in the loop-reversible state collisions between particles survive even in the steady state but the particles move in a reversible fashion with following non-affine loop trajectories. The simulations have predicted that the loop-reversible state intervenes between the point-reversible and irreversible states. It is noted that the loop-reversible state was not discussed by the original random organization model [8] because randomly kicked particles cannot form the closed loop trajectories. The loop-reversible behavior was later verified in a 2D soft-jammed colloidal system under global shear [31].

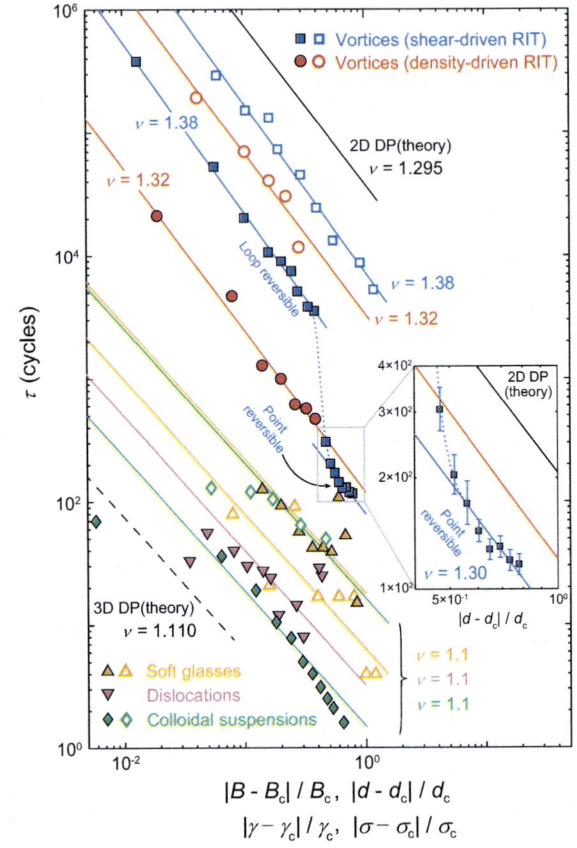

Fig. 3.5 The critical behavior of τ in different experimental systems. The log-log plots of τ versus the dimensionless control parameters that drive RIT. Each symbol indicates the data around the shear-driven RIT in colloidal suspensions (green diamonds) [8], soft glasses (yellow upright triangles) [22], and dislocations in small crystals (magenta inverted triangles) [23]. Our data around the density-driven RIT (red circles) and the shear-driven RIT for $f = 450$ kHz (blue squares) are plotted together. The quantities, $|B - B_c|/B_c$, $|d - d_c|/d_c$, $|\gamma - \gamma_c|/\gamma_c$, and $|\sigma - \sigma_c|/\sigma_c$, on the horizontal axis are the normalized distance from the critical point, where γ represents the strain amplitude for colloidal suspensions and soft glasses, and σ stands for the stress for dislocations. The data shown with closed and open symbols are acquired in the reversible and irreversible states, respectively. Colored solid lines are the fits of τ to the power of each control parameter with exponent ν, whose value is displayed in the main panel. The values of critical point are $B_c = 3.26$, $d_c = 45.2$ nm, $\gamma_c = 1.71$ [8], $\gamma_c = 0.25$ [22], and $\sigma_c = 390$ MPa [23]. Black solid and dashed lines have slopes of $\nu = 1.295$ and 1.110 expected for the DP universality class in 2D and 3D, respectively [21, 25]. The blue dotted line is a guide to the eye. In the reversible phase of the vortex system, τ shown with closed blue squares drops significantly at d_{c1} (= 25 nm), indicating the transition or the crossover from the loop-reversible to point-reversible state. Inset: The enlarged view of the square area in the main panel. τ deep inside the reversible phase of the shear-driven RIT for the vortex system is plotted against ($|d - d_c|/d_c$) in a double logarithmic scale. The same symbols and lines are used as those in the main panel. The solid blue line shows the fit of power law, $\tau \propto |d - d_c|^{-\nu}$, with $\nu = 1.30 \pm 0.18$ in the point-reversible regime ($d < d_{c1}$). Reprinted from Ref. [2] by S. Maegochi et al. licensed under CC BY 4.0

We consider that these two reversible states are detected in our experiments. Here, we define the threshold particle numbers, $N_c \equiv d_c/a_0$ and $N_c \equiv d/a_{0c}$, at the critical point of the shear-driven and density-driven RIT, respectively. We find $N_c = 45\ \mathrm{nm}/26\ \mathrm{nm} = 1.7$ for the shear-driven RIT and $N_c = 60\ \mathrm{nm}/27\ \mathrm{nm} = 2.2$ for the density-driven RIT, which are about 2. Therefore, our vortex system undergoes RIT when the vortices move about twice the intervortex spacing. This indicates that near the RIT ($d_{c1} < d < d_c$), the collisions between vortices still survive even in the reversible steady state, which should correspond to the loop-reversible state. On the other hand, for $d < d_{c1} (\approx a_0)$, the vortex-vortex collisions are suppressed and the random organization less likely occurs, which is reflected in the significant drop in $\tau(d)$. Accordingly, this regime below d_{c1} corresponds to the point-reversible state. The similar drop in τ within the small strain regime has been reported by the recent simulation [6]. Figure S11 of Ref. [6] shows the transition or the crossover from the loop-reversible state with non-affine trajectories to the point-reversible state with affine trajectories as the amplitude of strain is decreased.

As shown with closed red circles in Fig. 3.5, $\tau(B)$ in the density-driven experiment does not show the drop with decreasing B. This is merely because the range of B in our experiment is limited to the peak effect regime, $B = 2.0$–4.2 T, in which pinning is effective and random organization is more pronounced. Particularly, the fixed value of $d = 60$ nm in this work is always larger than $a_0 = 24$–35 nm at $B = 4.2$–2.0 T. For investigating the point-reversible regime, the field B must be lower than 0.65 T which meets the condition of $a_0(B) > d\ (= 60\ \mathrm{nm})$. However, it is difficult to clearly observe the voltage relaxation in the low-B regime where pinning that causes local shear is not effective. When we fix d smaller than 60 nm instead, we are not able to explore the irreversible regime that exists above about $2a_0$ because a_0 in this work is limited above 24 nm, corresponding to $B \leq 4.2$ T.

We expect that the two types of reversible states and the transition or the crossover between them observed here will be also observed in other particlelike systems. In a dilute particlelike system of colloidal suspensions [3, 8], the collisionless point-reversible state has been found to extend over a wide range of the reversible regime. The loop-reversible state may possibly lie in the vicinity of the critical point. In a relatively dense particlelike system of soft glasses [22], on the other hand, the loop-reversible behavior has been observed over a wide range of the reversible regime, where the point-reversible state is possibly realized deep inside the reversible regime. The sheared soft glasses in the loop-reversible state involve a number of local plastic deformation, whose correlations in space grow on approaching the yielding transition (or RIT) [22, 32, 33]. The connection between the yielding transition and RIT has been suggested by simulations [6, 7, 34–39].

3.4 Conclusion

We studied the critical behavior of RIT driven by the vortex density B, as well as the shear amplitude d. We observed the critical behavior of the density-driven RIT manifested as the power-law divergence of $\tau(B)$ from both sides of the transition and obtained two critical exponents $\alpha = 0.45 \pm 0.05$ and $\nu = 1.32 \pm 0.10$. These values are, within errors, close to $\alpha = 0.40 \pm 0.05$ and $\nu = 1.38 \pm 0.08$ obtained for the shear-driven RIT [17]. The results show the universality of RIT independent of the control parameters B and d, and the universality class of both types of RIT turns out to be the same as 2D DP, which predicts $\alpha \approx 0.45$ and $\nu = 1.295 \pm 0.006$ [21].

We also found the transition or the crossover between two distinct reversible states in the shear-driven experiment, corresponding to the collisionless point-reversible state $d < d_{c1} (\approx a_0)$ and the loop-reversible state $(d_{c1} < d < d_c)$ with vortex-vortex collisions. In either reversible regime, $\tau(d)$ exhibits a power-law divergence at the same d_c with the almost identical exponent: $\nu = 1.38$ for the loop-reversible state and 1.30 for the point-reversible state.

References

1. Reichhardt C, Regev I, Dahmen K, Okuma S, Reichhardt CJO (2023) Reversible to irreversible transitions in periodic driven many-body systems and future directions for classical and quantum systems. Phys Rev Res 5:021001
2. Maegochi S, Ienaga K, Okuma S (2021) Critical behavior of density-driven and shear-driven reversible-irreversible transitions in cyclically sheared vortices. Sci Rep 11:19280
3. Pine DJ, Gollub JP, Brady JF, Leshansky AM (2005) Chaos and threshold for irreversibility in sheared suspensions. Nature 438:997
4. Pham P, Butler JE, Metzger B (2016) Origin of critical strain amplitude in periodically sheared suspensions. Phys Rev Fluids 1:022201(R)
5. Schreck CF, Hoy RS, Shattuck MD, O'Hern CS (2013) Particle-scale reversibility in athermal particulate media below jamming. Phys Rev E 88:052205
6. Das P, Vinutha HA, Sastry S (2020) Unified phase diagram of reversible-irreversible, jamming, and yielding transitions in cyclically sheared soft-sphere packings. Proc Natl Acad Sci USA 117:10203
7. Matsuyama H, Toyoda M, Kurahashi T, Ikeda A, Kawasaki T, Miyazaki K (2021) Geometrical properties of mechanically annealed systems near the jamming transition. Eur Phys J E 44:133
8. Corté L, Chaikin PM, Gollub JP, Pine DJ (2008) Random organization in periodically driven systems. Nat Phys 4:420
9. Brown BL, Reichhardt C, Reichhardt CJO (2019) Reversible to irreversible transitions in periodically driven skyrmion systems. New J Phys 21:013001
10. Milz L, Schmiedeberg M (2013) Connecting the random organization transition and jamming within a unifying model system. Phys Rev E 88:062308
11. Tjhung E, Berthier L (2015) Hyperuniform density fluctuations and diverging dynamic correlations in periodically driven colloidal suspensions. Phys Rev Lett 114:148301
12. Marković N, Christiansen C, Goldman AM (1998) Thickness-magnetic field phase diagram at the superconductor-insulator transition in 2D. Phys Rev Lett 81:5217
13. Yamanouchi M, Ieda J, Matsukura F, Barnes SE, Maekawa S, Ohno H (2007) Universality classes for domain wall motion in the ferromagnetic semiconductor (Ga, Mn)As. Science 317:1726
14. Wilken S, Guerra RE, Pine DJ, Chaikin PM (2020) Hyperuniform structures formed by shearing colloidal suspensions. Phys Rev Lett 125:148001
15. Jeanneret R, Bartolo D (2014) Geometrically protected reversibility in hydrodynamic Loschmidt-echo experiments. Nat Commun 5:3474
16. Weijs JH, Jeanneret R, Dreyfus R, Bartolo D (2015) Emergent hyperuniformity in periodically driven emulsions. Phys Rev Lett 115:108301

17. Maegochi S, Ienaga K, Kaneko S, Okuma S (2019) Critical behavior near the reversible-irreversible transition in periodically driven vortices under random local shear. Sci Rep 9:16447
18. Adhikari M, Sastry S (2018) Memory formation in cyclically deformed amorphous solids and sphere assemblies. Eur Phys J E 41:105
19. Miyagawa K, Maegochi S, Ienaga K, Kaneko S, Okuma S (2021) Effects of the velocity on the reversible-irreversible transition in a periodically sheared vortex system. J Phys: Conf Ser 1975:012002
20. Reichhardt C, Reichhardt CJO (2009) Random organization and plastic depinning. Phys Rev Lett 103:168301
21. Hinrichsen H (2000) Non-equilibrium critical phenomena and phase transitions into absorbing states. Adv Phys 49:815
22. Nagamanasa KH, Gokhale S, Sood AK, Ganapathy R (2014) Experimental signatures of a nonequilibrium phase transition governing the yielding of a soft glass. Phys Rev E 89:062308
23. Ni X, Zhang H, Liarte DB, McFaul LW, Dahmen KA, Sethna JP, Greer JR (2019) Yield precursor dislocation avalanches in small crystals: the irreversibility transition. Phys Rev Lett 123:035501
24. Zhou C, Reichhardt CJO, Reichhardt C, Beyerlein I (2014) Random organization in periodically driven gliding dislocations. Phys Lett A 378:1675
25. Lübeck S (2004) Universal scaling behavior of non-equilibrium phase transitions. Int J Mod Phys B 18:3977
26. Menon GI, Ramaswamy S (2009) Universality class of the reversible-irreversible transition in sheared suspensions. Phys Rev E 79:061108
27. Franceschini A, Filippidi E, Guazzelli E, Pine DJ (2011) Transverse alignment of fibers in a periodically sheared suspension: an absorbing phase transition with a slowly varying control parameter. Phys Rev Lett 107:250603
28. Fisher ME (1967) The theory of equilibrium critical phenomena. Rep Prog Phys 30:615
29. Kadanoff LP, Götze W, Hamblen D, Hecht R, Lewis EAS, Palciauskas VV, Rayl M, Swift J, Aspnes D, Kane J (1967) Static phenomena near critical points: theory and experiment. Rev Mod Phys 39:395
30. Nagasawa K, Miyazaki K, Kawasaki T (2019) Classification of the reversible-irreversible transitions in particle trajectories across the jamming transition point. Soft Matter 15:7557
31. Keim NC, Arratia PE (2014) Mechanical and microscopic properties of the reversible plastic regime in a 2D jammed material. Phys Rev Lett 112:028302
32. Martens K, Bocquet L, Barrat J-L (2011) Connecting diffusion and dynamical heterogeneities in actively deformed amorphous systems. Phys Rev Lett 106:156001
33. Ghosh A, Budrikis Z, Chikkadi V, Sellerio AL, Zapperi S, Schall P (2017) Direct observation of percolation in the yielding transition of colloidal glasses. Phys Rev Lett 118:148001
34. Fiocco D, Foffi G, Sastry S (2013) Oscillatory athermal quasistatic deformation of a model glass. Phys Rev E 88:020301(R)
35. Regev I, Weber J, Reichhardt C, Dahmen KA, Lookman T (2015) Reversibility and criticality in amorphous solids. Nat Commun 6:8805
36. Kawasaki T, Berthier L (2016) Macroscopic yielding in jammed solids is accompanied by a nonequilibrium first-order transition in particle trajectories. Phys Rev E 94:022615
37. Leishangthem P, Parmar ADS, Sastry S (2017) The yielding transition in amorphous solids under oscillatory shear deformation. Nat Commun 8:14653
38. Szulc A, Gat O, Regev I (2020) Forced deterministic dynamics on a random energy landscape: implications for the physics of amorphous solids. Phys Rev E 101:052616
39. Otsuki M, Hayakawa H (2022) Softening and residual loss modulus of jammed grains under oscillatory shear in an absorbing state. Phys Rev Lett 128:208002

Chapter 4
Moving Smectic Phase and Transverse Mode Locking in Driven Vortex Matter

4.1 Introduction

As introduced in Sect. 1.1, condensed matter systems out of equilibrium exhibit various ordering phenomena, where how the disordered system reorders dynamically is a general question. As a typical example of the ordering phenomena, hereafter, we consider the dynamical ordering of vortices driven over the disordered landscape explained in Sect. 1.1.3. In the vortex system, the interaction between vortices and random pinning centers on a substrate is reduced by increasing the applied current I (driving force) and vortices successively recover its configurational order. With increasing I, initially pinned vortices are set in the disordered plastic flow just above a depinning current I_{d} and considered to gradually transform into the anisotropically ordered moving smectic phase [1–6]. They eventually reorder into the isotropic moving Bragg glass with sixfold orientational order.

As already shown in Fig. 1.10, the smectic flow is described as the anisotropic structure with periodicity transverse to the driving direction and is considered to be generic to driven 2D systems [1, 7] such as moving Wigner crystals [8], charge transport in disordered arrays of metallic dots [9], moving skyrmions [10], and flowing emulsions [11]. Contrary to many theoretical works, there is still no experimental evidence for the moving smectic phase in current-induced dynamical ordering of vortices. The smectic structure of vortices driven by a field gradient was detected by a Bitter decoration [12, 13], while the decoration technique has limitations in the range of B below several tens of millitesla and average velocity of vortices below $v \approx 1$ µm/s, corresponding to a slowly creeping regime. In addition, the systematic change of the driving force was not carried out. Therefore, the dynamical ordering from the plastic flow to smectic flow to moving Bragg glass state have not been elucidated by any experimental studies.

In this chapter, to obtain the dynamical phase diagram of current-driven vortices, we develop two-step measurements constructed from mutually perpendicular drives [14] and a transverse mode-locking (ML) technique as described below. Using these methods, we observe the current-induced dynamical orderings successively from the

S. Maegochi, *Nonequilibrium Phase Transitions in Driven Vortex Matter*, Springer Theses,
https://doi.org/10.1007/978-981-97-2987-6_4

plastic flow to anisotropic smectic flow to isotropic moving Bragg glass, together with convincing evidence of the moving smectic phase in the high velocity regime ($v \approx 1$ m/s) [15].

4.2 Experiments

We used the cross-shaped a-Mo_xGe_{1-x} film with weak random pinning, presented in Sect. 2.2. Figure 4.1a shows schematics of the sample design with current ($\pm I_x, \pm I_y$) and voltage ($\pm V_x, \pm V_y$) electrodes. The distance between corresponding voltage electrodes is $l = 1.95$ mm. We performed standard four probe measurements in the x direction using the ($\pm I_x, \pm V_x$) electrodes and in the y direction using the ($\pm I_y, \pm V_y$) electrodes. To generate vortices, we applied B normal to the film surface. All the experiments were conducted at 3.6 K and 1.0 T, corresponding to the Bragg glass phase at equilibrium. The size of the vortex core ξ and the interaction range of vortices λ are about $\approx 1 \times 10$ and $\approx 10^2$ nm, respectively [16].

4.2.1 Two-Step Measurements

In the two-step measurements, we measure the time evolution of voltage $V(t)$ immediately after applying a square current within the rise time of 0.5 μs using the oscilloscope. The steady-state voltage is defined as $V^{\infty}(\equiv V(t \to \infty))$, where t stands for the real time, not the cycle number used in the previous chapters. Figure 4.1b shows the experimental procedure. For initialization, we applied a large current about 20 mA, corresponding to the moving Bragg glass state as identified below, for a long time and prepared the highly ordered state to erase the possible history effect. In other words, we set the system to the stable minimum in the PEL without any memories. Then, for the first step, the current I with a given magnitude is applied in the minus x direction to drive the vortices in the y direction as shown in Fig. 4.1c and is turned off in the steady state to freeze the final vortex configuration. Next, for the second step, we applied the same drive (I) in the same direction as the first one, which we call a parallel second drive, Drive y. We also performed the second-step measurements shown in Fig. 4.1d, where we applied a drive rotated by 90° clockwise relative to the first one. We call this step a perpendicular second drive, Drive x. This two-step protocol is an alteration of two-step measurements using successive parallel drives as mentioned in Sect. 1.2.2. From the second-step measurements, we can know the degree of order of steady-state vortex configurations in the first step via the memory effect.

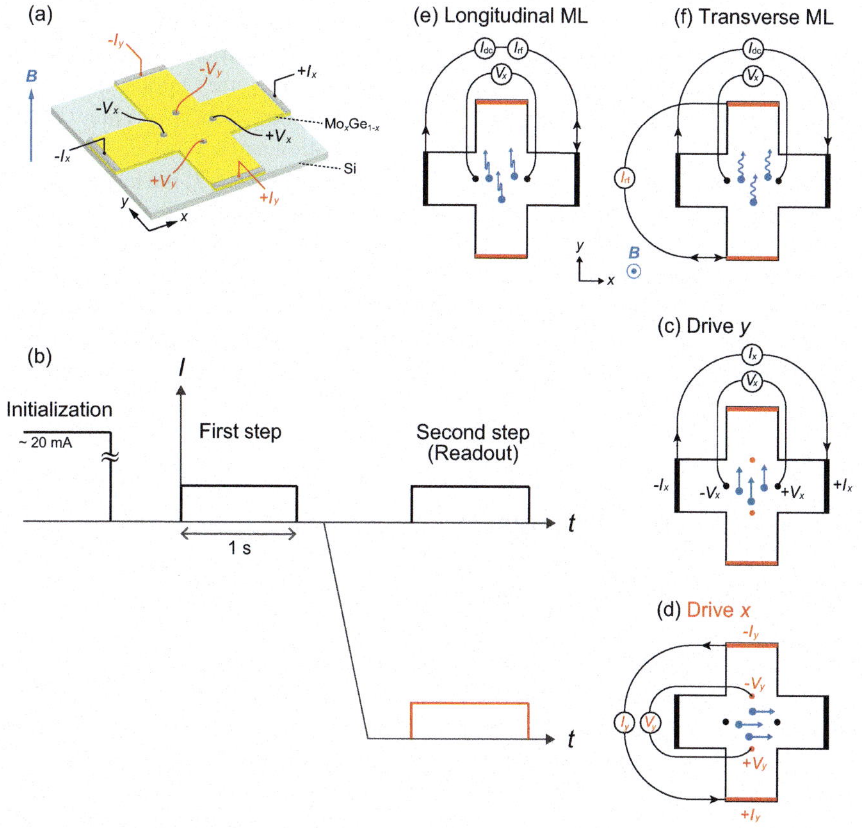

Fig. 4.1 **a** Schematic illustration of the cross-shaped a-$\mathrm{Mo}_x\mathrm{Ge}_{1-x}$ film and with electrodes. The magnetic field B is applied normal to the plane of the film. **b** The experimental procedure for the two-step measurements. **c**–**f** The experimental setup. **c** The first-step measurement and the second-step measurement of parallel drive (Drive y). **d** The second-step measurement of perpendicular drive (Drive x). **e** The longitudinal ML measurement. **f** The transverse ML measurement. Blue dots and arrows represent the schematic vortex motion by the Lorentz-like driving force

4.2.2 *Mode-Locking Measurements*

ML is a dynamic interference phenomenon observed in periodic media driven over ordered or disordered substrates such as charge-density waves [17–22], Josephson junction arrays [23–25], colloids [26], skyrmions [27], and vortices in type-II superconductors [28–40]. For instance, when the vortex lattice moves at constant velocity v by the dc drive, the disorder on the substrate produces a periodic modulation of the vortex motion. The frequency of this modulation, specifically called the internal frequency f_{int}, is given by $f_{\text{int}} = v/a$, where a is the lattice period in the dc flow direction. When the rf drive with the external frequency f_{ext} is superimposed on the dc drive and f_{ext} matches f_{int} itself or its harmonics, two frequencies are locked as

$pf_{\rm int} = qf_{\rm ext}$ with integers (p, q). As a result, ML steps akin to Shapiro steps [41] appear in the I–V curves and also peak structures in the differential conductance dI/dV are observed at $V_{p/q} = (p/q) f_{\rm ext} aBl$. Since the vortex lattice can take its principal axis parallel and perpendicular to the moving direction, two values of a are possible [29, 37, 38, 42]: $a = a_0$ for parallel orientation and $a = \sqrt{3}a_0$ for perpendicular orientation, where $a_0 = (2\Phi_0/\sqrt{3}B)^{1/2}$ (same as Eq. 1.13) is a lattice spacing ($\approx$ 50 nm at 1.0 T). Through ML experiments, one can obtain the microscopic period of fast driven media by the macroscopic force-velocity measurements. In conventional ML experiments, the rf drive was always superimposed in the same direction as the dc drive (longitudinal ML) as shown in Fig. 4.1e. To detect the transverse order of the moving smectic phase, here, we develop the transverse ML [43], where the rf drive is superimposed in the orthogonal direction to the dc drive as shown in Fig. 4.1f.

4.3 Results and Discussion

4.3.1 Two-Step Measurements

Figure 4.2b shows the I–V characteristics, where voltage appears at the depinning current $I_{\rm d} = 4.7$ mA defined using a 10^{-8} V criterion. The I–V curve has the upward curvature above $I_{\rm d}$, indicating that vortices are set in the disordered plastic flow. Figure 4.3a shows voltage responses $V(t)$ to Drive y with the black line and to Drive x with the red line, where the driving current yielding $V^{\infty} = 0.01$ mV is applied. For Drive y, $V(t)$ rises sharply and reaches V^{∞} without showing the relaxation. For Drive x, a initial rise of $V(t)$ exceeds V^{∞} and the decreasing relaxation is observed toward V^{∞}. At $t \approx 0$ of Drive x, the vortices move easier compared to them in the final state of the first-step measurement, while the vortex flow under Drive y is the same as that in the first step.

This anisotropy of the responses to the second drives indicate that the frozen vortices retain a memory of the direction of the former drive applied in the first step. This result is described as directional fragility of the clogging phenomena [14, 44–47]. When the particles flow through disordered media with walls and constrictions, their transport is interrupted by the formation of clogs. The clogging is heterogeneous and anisotropic phenomenon and hence easily eliminated by switching the direction of the drive. We conducted the measurements shown in Fig. 4.3a at the low velocity regime just above $I_{\rm d}$, where the vortex system is in the disordered plastic flow state involving many pinned vortices. Most of the pinned vortices are apart from each other because of the random distribution of the pinning centers in our film, but some vortices are pinned nearby and can construct wall-like structures [44–47]. As shown with red dots in Fig. 4.3b, a part of flowing vortices is caught by the walls at the end of the first-step measurement. Under Drive y of the second-step measurement, these vortices are still caught by the walls and the flow state does not change, so that $V(t)$

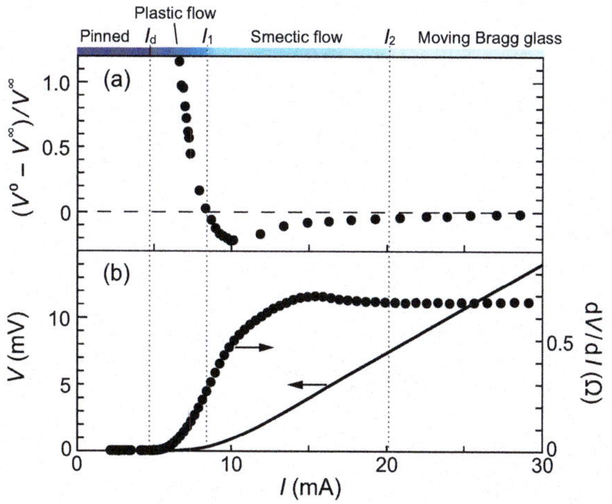

Fig. 4.2 **a** I versus $(V^0 - V^\infty)/V^\infty$ obtained from $V(t)$ in the second-step measurements of Drive x, where V^0 and V^∞ denote initial and steady-state voltages, respectively. The horizontal dashed line marks the position of zero. The dynamical phase diagram is demonstrated at the top. **b** I-V characteristics and corresponding $\mathrm{d}V/\mathrm{d}I$ shown with solid line and circles, respectively. From left to right, vertical dotted lines indicate the position of characteristic currents: I_d, I_1 at which a sign of $(V^0 - V^\infty)/V^\infty$ is reversed, and I_2 above which $\mathrm{d}V/\mathrm{d}I$ takes a constant value. Reprinted from Ref. [15] by S. Maegochi et al. licensed under CC BY 4.0

exhibits no relaxation. On the other hand, when the direction of the drive is switched by applying Drive x, some vortices are released from the walls and start to flow as shown with red arrows in Fig. 4.3c, resulting in a large initial rise of $V(t)$. These flowing vortices are subsequently trapped by other walls again as drawn by red dots in Fig. 4.3d. Thus, $V(t)$ for Drive x in Fig. 4.3a exhibits a monotonic decrease to V^∞.

We also conduct the same two-step measurements for larger driving currents yielding larger V^∞. In Fig. 4.4a, the normalized voltage responses $V(t)/V^\infty$ to Drive x of the second-step measurements are plotted, where the values of V^∞ are 0.1, 0.4, and 0.7 mV from top to bottom. For $V^\infty = 0.1$ mV, $V(t)$ exhibits a similar decreasing relaxation to that for $V^\infty = 0.01$ mV shown with the red line in Fig. 4.3a. With increasing V^∞, a decreasing relaxation of $V(t)$ is suppressed substantially for $V^\infty = 0.4$ mV, and turns into a monotonically increasing relaxation for $V^\infty = 0.7$ mV. The results indicate that the plastic flow with clogging that is the origin of the decreasing relaxation still survives for $V^\infty = 0.1$ mV. As V^∞ is increased, the number of pinned vortices decreases, leading to a small number of walls and suppression of the clogging. For $V^\infty = 0.7$ mV, the clogging phenomenon observed in the plastic flow regime is subject to or replaced by another phenomenon that explains the increasing relaxation. Since $V(t)$ exhibits the increasing relaxation in the relatively high velocity regime ($I > 8.4$ mA), we consider its origin as the directional robustness of the smectic flow.

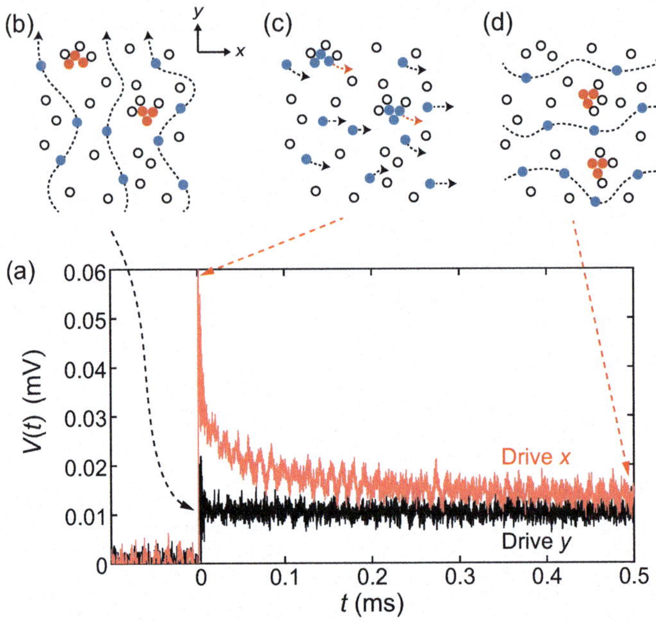

Fig. 4.3 Directional fragility of the clogging. **a** Voltage responses $V(t)$ in second-step measurements for Drive y (black) and Drive x (red), where the first drive was applied in the y direction. **b–d** Schematic illustrations of vortex flow at the beginning of Drive y (b), and at the beginning **c** and at the end **d** of Drive x. The pinned vortices are drawn as open circles. Blue and red dots represent the flowing and trapped vortices, respectively. Reprinted from Ref. [15] by S. Maegochi et al. licensed under CC BY 4.0

Figures 4.4b–d show schematic description of the results. As shown in Fig. 4.4b, at the end of the first-step measurements in the smectic regime, the flowing vortices form 1D channels in the driving direction y. These smectic channels possess the periodicity in the transverse direction, namely, in the x direction, but there is no positional order in the longitudinal direction y. When these vortices are subjected to the perpendicular second drive (Drive x), they cannot flow in straight channels in the x direction immediately, rather in winding channels as illustrated in Fig. 4.4c, which gives lower mobility and a small initial rise of $V(t)$. As rearrangements of the vortices proceed, the vortices self-organize into the 1D channels along the x direction as shown in Fig. 4.4d, resulting in monotonically increasing relaxation of $V(t)$. The monotonic relaxation curves for $V^{\infty} = 0.1$ and 0.7 mV are well fitted by a single exponential function as shown by black solid lines in Fig. 4.4a. Meanwhile, the relaxation curve for $V^{\infty} = 0.4$ mV is nonmonotonic and fitted by the function with two exponential terms as shown by a dotted line, implying that the clogging and smectic mechanisms are coexistent. We also find in the inset of Fig. 4.4a that $V(t)$ for sufficiently large drive, $V^{\infty} = 10$ mV, shows almost no relaxation. This implies that the anisotropic smectic flow reorders into the isotropic moving Bragg glass.

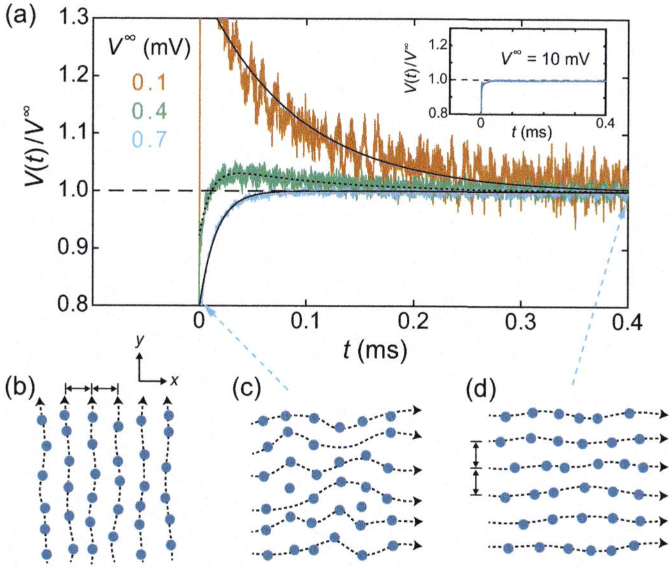

Fig. 4.4 Directional robustness of the smectic flow. **a** Voltage responses $V(t)/V^{\infty}$ in second-step measurements for Drive x yielding $V^{\infty} = 0.1, 0.4$, and 0.7 mV from top to bottom. Solid and dotted lines are the fitting of the data explained in the text. The horizontal dashed line indicates the location of the steady-state value, $V(t)/V^{\infty} = 1$. Inset: $V(t)/V^{\infty}$ subjected to Drive x with $V^{\infty} = 10$ mV. **b** The final vortex configuration for the first-step measurement in the smectic flow regime. Dotted lines show the smectic-flow channels with periodicity in the x direction. **c, d** Schematic illustrations of vortex flow in the second-step measurement. **c** Just after the application of Drive x to the frozen configuration in **b**. **d** In the steady state of Drive x. The vortices form the smectic-flow channels with periodicity in the y direction. Reprinted from Ref. [15] by S. Maegochi et al. licensed under CC BY 4.0

To see continuous changes in flow states, in Fig. 4.2a, we plot $(V^0 - V^{\infty})/V^{\infty}$ for Drive x with various amplitudes against I. Here, $V^0[\equiv V(t = 2\ \mu\text{s})]$ represents the initial voltage at $t = 2\ \mu$s, which corresponds to the response time of the amplifier. A sign of $(V^0 - V^{\infty})/V^{\infty}$ changes from positive to negative at 8.4 mA($\equiv I_1$). This indicates the transition or the crossover from the plastic flow with clogging to the smectic flow around I_1. After taking a minimum, $(V^0 - V^{\infty})/V^{\infty}$ exhibits a gradual increase and convergence toward zero, marked by a horizontal dashed line, above $I \approx 20$ mA. This value is very close to 20 mA($\equiv I_2$) above which the differential resistance $\mathrm{d}V/\mathrm{d}I$ in Fig. 4.2b takes a constant value. I_2 is usually interpreted as an onset of the moving Bragg glass state [4, 48–50]. Thus, at the top of Fig. 4.2a, we show the dynamical phase diagram of vortices constructed from the two-step measurements. The flow state of vortices changes from the plastic flow to smectic flow around I_1, and from the smectic flow to moving Bragg glass around I_2. We note that whether these changes of flow states are phase transitions or crossovers are unknown at this time. In the next chapter, we will reveal that the change between the plastic and smectic flows is the second-order phase transition.

4.3.2 Transverse and Longitudinal Mode Locking

In order to detect the smectic order of flowing vortices more directly, we perform the ML measurements. Figure 4.5a shows $\mathrm{d}I/\mathrm{d}V$ of conventional longitudinal ML, where the 25 MHz rf current I_{rf} with various amplitudes, 9.18, 9.86, and 10.54 mA from bottom to top, is superimposed on the dc current I in the same x direction. Therefore, vortices feel both the dc and ac driving forces in the y direction. In this work, for unspecified reasons, $f_{\mathrm{ext}} \approx 25$ MHz is the maximum frequency available for the ML measurements. Further investigation of the high-frequency region is a future work. The ML peaks appear at $V \approx 2$ mV in the smectic flow regime determined from the two-step measurements. This value is close to $V_{1/2}$ with $a = \sqrt{3}a_0$, as indicated by a vertical solid line, corresponding to the perpendicular orientation, as drawn in the inset. In Fig. 4.5b, $\mathrm{d}I/\mathrm{d}V$ of transverse ML is displayed, where the 25 MHz I_{rf} with various amplitudes, 9.24, 9.90, and 10.56 mA from bottom to top, is superimposed in the y direction orthogonal to the dc current. To our knowledge, this is the first experimental demonstration of transverse ML. Unlike longitudinal ML, we find the fundamental ML peaks at $V \approx 2.5$ mV, which is again within the smectic flow regime and is close to $V_{1/1}$ with $a = a_0$, as indicated by a vertical dashed line. This corresponds to the parallel orientation as drawn in the inset.

We have previously found that the orientation of the vortex lattice rotates from the perpendicular orientation to the parallel one with increasing the effective (total) velocity of vortices [38]. The velocity-induced reorientation of the vortex lattice takes place when a characteristic time τ for the vortex to travel a_0 is decreased down to $\tau_{\mathrm{th}} \approx 10$ ns. In the present work, the result on longitudinal ML meets this condition, $\tau(\approx 25 \text{ ns}) > \tau_{\mathrm{th}}$, which agrees with the perpendicular orientation we observed. For transverse ML, the vortex lattice takes parallel orientation, while the total velocity in the dc driving direction is not larger than that of longitudinal ML since dc and ac components of the velocity are orthogonal. Thus, the condition of $\tau > \tau_{\mathrm{th}}$ is not satisfied and the parallel orientation observed here does not result from the velocity-induced reorientation. Here, we find that for both ML experiments, as seen in the inset to Fig. 4.5a, b, the vortex lattices are in the perpendicular orientation with respect to the ac driving direction. Simulation has suggested that the vortex lattice favors the orientation where its principal lattice vector is aligned perpendicular to the direction of oscillating motion [42], which agrees with the present results of both transverse and longitudinal ML.

In Fig. 4.5, the most striking feature is the difference in the height of ML peaks. Transverse ML shows considerably larger ML peaks than longitudinal ML. This indicates that vortices in the transverse ML state move more coherently than those in the longitudinal ML state. This result is well described by the fact that the transverse order is higher than the longitudinal one in the moving smectic phase. Several simulations [3, 4, 6] have shown that the smectic flow under the dc drive contains some aligned dislocations between adjacent uncoupled channels. Their Burgers vectors are directed in the driving force, so that dislocations glide only in the direction of the driving force. Once the dislocations glide, adjacent uncoupled flow channels

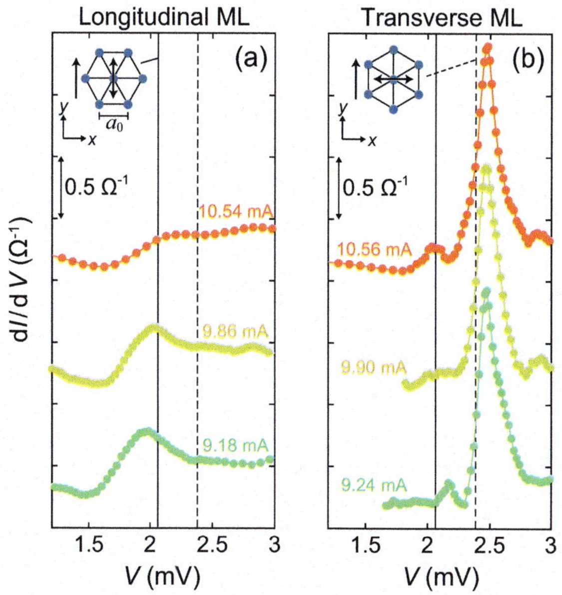

Fig. 4.5 **a** $\mathrm{d}I/\mathrm{d}V$ versus V for longitudinal ML. **b** $\mathrm{d}I/\mathrm{d}V$ versus V for transverse ML. In each figure, the amplitudes of 25 MHz rf current superimposed on the dc current, I_{rf}, are displayed. For the sake of clarity, curves are shifted in the vertical direction. Vertical solid and dashed lines respectively show the calculated positions of ML peaks for the perpendicular and parallel orientations. The schematic vortex lattices with perpendicular and parallel orientations are drawn in the insets of **a** and **b**, respectively. The flow directions by dc and ac drives are shown by single and double arrows, respectively. a_0 is the lattice spacing. Reprinted from Ref. [15] by S. Maegochi et al. licensed under CC BY 4.0

easily slip past each other, which leads to different velocity v between these channels. The presence of dislocations also indicates that adjacent uncoupled channels contain different numbers of vortices, and thus the lattice period a differs between these channels [6]. This difference in v and a results in different ML conditions between the uncoupled channels.

When we superimpose the longitudinal ac drive on the dc drive, the phases of the vortex motion between some coupled channels are locked. However, the phases between uncoupled channels are not locked simultaneously due to the above reason. This accounts for the relatively small peaks of longitudinal ML in Fig. 4.5a. In contrast, when we superimpose the transverse ac drive, all channels are forced to oscillate in the transverse direction in the same phase due to the transverse periodicity of the smectic-flow channels [1–6]. This accounts for the observation of the large transverse-ML peaks in Fig. 4.5b. Similar situation with geometrical constraints is seen in other elastic media such as charge-density waves, where their original periodicity in the longitudinal direction gives rise to a sharp longitudinal ML resonance [17–22]. It is noted that the periodicity in the charge-density wave is inherent, while that in the vortex system is induced dynamically. In this section, we have developed the transverse ML measurements and shown that both the transverse and longitudinal ML techniques are powerful and convenient tool to detect the anisotropic periodicity in fast driven media. In the present work, the ML techniques allowed us to directly observe the dynamically induced transverse periodicity of smectic flow in driven vortices with large velocity of $v \approx 1$ m/s, where methods for visualization are not available.

4.4 Conclusion

We investigated the dynamical phases of vortices driven over random pinning. We prepared the cross-shaped sample and developed the two-step measurements consisting of mutually perpendicular drives (I). Using this method, we found the current-induced dynamical orderings from the plastic flow to the anisotropic smectic flow to the isotropic moving Bragg glass. In the plastic flow regime, the directional fragility associated with clogging was observed. We also developed the transverse ML measurements. We observed the first transverse ML, which exhibits larger signals than longitudinal ML due to the higher transverse order of the smectic flow than the longitudinal one. The result provides convincing evidence of the moving smectic phase in between the plastic flow and the moving Bragg glass phase. The transverse ML technique developed here combined with longitudinal ML is a powerful and convenient method to detect the anisotropic structure in fast driven periodic media.

References

1. Balents L, Marchetti MC, Radzihovsky L (1998) Nonequilibrium steady states of driven periodic media. Phys Rev B 57:7705
2. Le Doussal P, Giamarchi T (1998) Moving glass theory of driven lattices with disorder. Phys Rev B 57:11356
3. Moon K, Scalettar RT, Zimányi GT (1996) Dynamical phases of driven vortex systems. Phys Rev Lett 77:2778
4. Olson CJ, Reichhardt C, Nori F (1998) Nonequilibrium dynamic phase diagram for vortex lattices. Phys Rev Lett 81:3757
5. Kolton AB, Domínguez D, Grønbech-Jensen N (1999) Hall noise and transverse freezing in driven vortex lattices. Phys Rev Lett 83:3061
6. Fangohr H, Cox DJ, de Groot PAJ (2001) Vortex dynamics in two-dimensional systems at high driving forces. Phys Rev B 64:064505
7. Reichhardt C, Reichhardt CJO (2017) Depinning and nonequilibrium dynamic phases of particle assemblies driven over random and ordered substrates: a review. Rep Prog Phys 80:026501
8. Reichhardt C, Olson CJ, Grønbech-Jensen N, Nori F (2001) Moving wigner glasses and smectics: dynamics of disordered wigner crystals. Phys Rev Lett 86:4354
9. Reichhardt C, Reichhardt CJO (2003) Charge transport transitions and scaling in disordered arrays of metallic dots. Phys Rev Lett 90:046802
10. Díaz SA, Reichhardt CJO, Arovas DP, Saxena A, Reichhardt C (2017) Fluctuations and noise signatures of driven magnetic skyrmions. Phys Rev B 96:085106

11. Blay ML, Adda-Bedia M, Bartolo D (2020) Emergence of scale-free smectic rivers and critical depinning in emulsions driven through disorder. Proc Natl Acad Sci USA 117:13914
12. Pardo F, de la Cruz F, Gammel PL, Bucher E, Bishop DJ (1998) Observation of smectic and moving-Bragg-glass phases in flowing vortex lattices. Nature 396:348
13. Marchevsky M, Aarts J, Kes PH (1999) Depinning and anisotropic order in flowing and static vortex lattices in $NbSe_2$ studied with magnetic decoration. Phys Rev B 60:14601
14. Reichhardt C, Reichhardt CJO (2020) Jamming, fragility and pinning phenomena in superconducting vortex systems. Sci Rep 10:11625
15. Maegochi S, Ienaga K, Okuma S (2022) Moving smectic phase and transverse mode locking in driven vortex matter. Phys Rev Res 4:033085
16. Nishio T, Okayasu S, Suzuki J, Kokubo N, Kadowaki K (2008) Observation of an extended magnetic field penetration in amorphous superconducting MoGe films. Phys Rev B 77:052503
17. Zettl A, Grüner G (1984) Phase coherence in the current-carrying charge-density-wave state: ac-dc coupling experiments in $NbSe_3$. Phys Rev B 29:755
18. Coppersmith SN, Littlewood PB (1986) Interference phenomena and mode locking in the model of deformable sliding charge-density waves. Phys Rev Lett 57:1927
19. Thorne RE, Tucker JR, Bardeen J (1987) Experiment versus the classical model of deformable charge-density waves: interference phenomena and mode locking. Phys Rev Lett 58:828
20. Middleton AA, Biham O, Littlewood PB, Sibani P (1992) Complete mode locking in models of charge-density waves. Phys Rev Lett 68:1586
21. Higgins MJ, Middleton AA, Bhattacharya S (1993) Scaling near mode locking in a charge density wave conductor. Phys Rev Lett 70:3784
22. Sinchenko AA, Monceau P, Crozes T (2012) Transverse conductivity in the sliding charge-density-wave state of $NbSe_3$. Phys Rev Lett 108:046402
23. Benz SP, Rzchowski MS, Tinkham M, Lobb CJ (1990) Fractional giant shapiro steps and spatially correlated phase motion in 2D josephson arrays. Phys Rev Lett 64:693
24. Lee KH, Stroud D, Chung JS (1990) Calculation of giant fractional shapiro steps in josephson-junction arrays. Phys Rev Lett 64:962
25. Domínguez D, José JV (1992) Giant shapiro steps with screening currents. Phys Rev Lett 69:514
26. Juniper MPN, Straube AV, Besseling R, Aarts DGAL, Dullens RPA (2015) Microscopic dynamics of synchronization in driven colloids. Nat Commun 6:7187
27. Sato T, Kikkawa A, Taguchi Y, Tokura Y, Kagawa F (2020) Mode locking phenomena of the current-induced skyrmion-lattice motion in microfabricated MnSi. Phys Rev B 102:180411(R)
28. Fiory AT (1971) Quantum interference effects of a moving vortex lattice in Al films. Phys Rev Lett 27:501
29. Schmid A, Hauger W (1973) On the theory of vortex motion in an inhomogeneous superconducting film. J Low Temp Phys 11:667
30. Harris JM, Ong NP, Gagnon R, Taillefer L (1995) Washboard frequency of the moving vortex lattice in $YBa_2Cu_3O_{6.93}$ detected by ac-dc interference. Phys Rev Lett 74:3684
31. Van Look L, Rosseel E, Van Bael MJ, Temst K, Moshchalkov VV, Bruynseraede Y (1999) Shapiro steps in a superconducting film with an antidot lattice. Phys Rev B 60:R6998
32. Kolton AB, Domínguez D, Grønbech-Jensen N (2001) Mode locking in ac-driven vortex lattices with random pinning. Phys Rev Lett 86:4112
33. Kokubo N, Besseling R, Vinokur VM, Kes PH (2002) Mode locking of vortex matter driven through mesoscopic channels. Phys Rev Lett 88:247004
34. Kokubo N, Kadowaki K, Takita K (2005) Peak effect and dynamic melting of vortex matter in $NbSe_2$ crystals. Phys Rev Lett 95:177005
35. Okuma S, Inoue J, Kokubo N (2007) Suppression of broadband noise at mode locking in driven vortex matter. Phys Rev B 76:172503
36. Okuma S, Imaizumi H, Kokubo N (2009) Intrinsic quantum melting of a driven vortex lattice in amorphous Mo_xGe_{1-x} films. Phys Rev B 80:132503
37. Okuma S, Imaizumi H, Shimamoto D, Kokubo N (2011) Quantum melting and lattice orientation of driven vortex matter. Phys Rev B 83:064520

38. Okuma S, Shimamoto D, Kokubo N (2012) Velocity-induced reorientation of a fast driven Abrikosov lattice. Phys Rev B 85:064508
39. Ochi A, Kawamura Y, Inoue T, Kaji T, Mihaly D, Kaneko S, Kokubo N, Okuma S (2016) Dynamic melting of driven Abrikosov lattices in an amorphous Mo_xGe_{1-x} film in tilted field. J Phys Soc Jpn 85:034712
40. Ochi A, Sohara N, Kaneko S, Kokubo N, Okuma S (2016) Equilibrium and dynamic vortex states near absolute zero in a weak pinning amorphous film. J Phys Soc Jpn 85:044701
41. Shapiro S (1963) Josephson currents in superconducting tunneling: the effect of microwaves and other observations. Phys Rev Lett 11:80
42. Valenzuela SO (2002) Order and mobility of solid vortex matter in oscillatory driving currents. Phys Rev Lett 88:247003
43. Kolton AB, Domínguez D, Grønbech-Jensen N (2002) Mode locking in driven vortex lattices with transverse ac drive and random pinning. Phys Rev B 65:184508
44. Reichhardt CJO, Groopman E, Nussinov Z, Reichhardt C (2012) Jamming in systems with quenched disorder. Phys Rev E 86:061301
45. Péter H, Libál A, Reichhardt C, Reichhardt CJO (2018) Crossover from jamming to clogging behaviors in heterogeneous environments. Sci Rep 8:10252
46. Reichhardt C, Reichhardt CJO (2018) Controlled fluidization, mobility, and clogging in obstacle arrays using periodic perturbations. Phys Rev Lett 121:068001
47. Stoop RL, Tierno P (2018) Clogging and jamming of colloidal monolayers driven across disordered landscapes. Commun Phys 1:68
48. Shi A-C, Berlinsky AJ (1991) Pinning and I-V characteristics of a two-dimensional defective flux-line lattice. Phys Rev Lett 67:1926
49. Bhattacharya S, Higgins MJ (1993) Dynamics of a disordered flux line lattice. Phys Rev Lett 70:2617
50. Xiao ZL, Andrei EY, Shuk P, Greenblatt M (2000) Equilibration and dynamic phase transitions of a driven vortex lattice. Phys Rev Lett 85:3265

Chapter 5
Evidence of Second-Order Transition and Critical Scaling for the Dynamical Ordering Transition

5.1 Introduction

In the previous chapter, we answered one of the long-standing questions whether the current-induced dynamical ordering from the plastic flow to the smectic flow actually takes place. We found dynamical ordering from the plastic flow to the anisotropic smectic flow as a function of the current using two-step measurements of transient voltage in response to mutually perpendicular drives. We also detect the moving smectic phase clearly by transverse ML. However, the central issue of whether the current-induced dynamical ordering is a phase transition or a crossover still remains to be elucidated. If it is revealed to be a true phase transition, it is also interesting to study whether its nature of the transition is first-order or second-order.

In this chapter, we develop transverse current-voltage (i.e., force-velocity) measurements to investigate the issue [1]. We measure the response of longitudinally moving vortices, instead of frozen vortices in the previous chapter, to the transverse drive using the same cross-shaped amorphous Mo_xGe_{1-x} film as used in the previous chapter. Then, we carry out the scaling analysis for the transverse current-voltage curves because the scaling approach is generally helpful to demonstrate the continuous second-order phase transition [2–4]. In fact, the critical scaling was employed to detect the second-order vortex-glass transition [5–17], while it has been applied only to the vortex state at equilibrium.

5.2 Experiments

We used the same cross-shaped amorphous Mo_xGe_{1-x} film with weak random pinning as used in the previous chapter and performed standard four probe measurements. Figure 5.1 shows the schematic experimental setup with arrangements of current (I_x, I_y) and voltage (V_y) electrodes used in the transverse current-voltage measurements. The vortices were generated by applying the magnetic field of 1.0

S. Maegochi, *Nonequilibrium Phase Transitions in Driven Vortex Matter*, Springer Theses,
https://doi.org/10.1007/978-981-97-2987-6_5

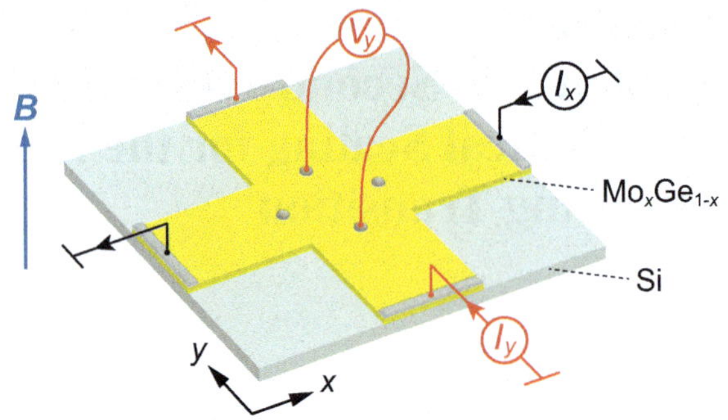

Fig. 5.1 Schematics of the experimental setup of the cross-shaped amorphous Mo_xGe_{1-x} film on the Si substrate. The magnetic field B is applied perpendicular to the film surface. Reprinted from Ref. [1] by S. Maegochi et al. licensed under CC BY 4.0

T normal to the surface of the film at 3.6 K, which is the same condition as in the previous chapter.

In this chapter, we mainly use the current density J and the electric field E in place of I and V, respectively. The depinning current densities in both x and y directions are $J_d = 0.86 \times 10^7$ A/m^2 (i.e., $I_d \approx 4.8$ mA). For the transverse current-voltage ($J_y - E_y$) measurements, we first applied the current in the minus x direction with a given current density J_x, where the drive was applied in the y direction. Then, we superimposed the transverse current in the y direction with density of J_y on J_x and examined the response of vortices to the transverse driving force exerted in the x direction by measuring the transverse $J_y - E_y$ characteristics in the y direction, as shown with red color in Fig. 5.1. In the transverse $J_y - E_y$ measurements, we recorded both $E_y(J_x, J_y)$ and $E_y(J_x, J_y = 0)$ for each J_y. Accordingly, the background signal, together with the small component of E_x coming from the possible misalignment of voltage electrodes, are safely subtracted and we can obtain reliable values of E_y.

5.3 Results and Discussion

5.3.1 *Transverse Depinning*

Figure 5.2 shows log-log plots of $J_y - E_y$ measured under various J_x. We note that J_x is the control parameter that alters the vortex flow state, while the transverse $J_y - E_y$ characteristics are used as a probe of the transverse vortex response. For smaller J_x, the $J_y - E_y$ curves have a negative curvature. With decreasing J_y, E_y decreases rapidly down to the background level of 10^{-8} V. Around $J_x \approx 1.38 \times 10^7$ A/m^2 ($\equiv J_x^*$), the $J_y - E_y$ curve takes a power-law functional form as indicated by a dashed straight line. For larger J_x above J_x^*, the $J_y - E_y$ curves have a positive curvature and E_y decreases linearly with decreasing J_y.

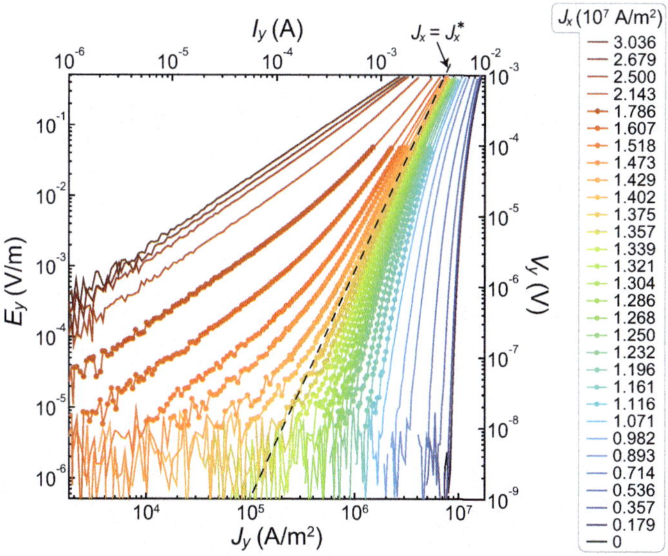

Fig. 5.2 Transverse current-voltage ($J_y - E_y$) characteristics measured under fixed J_x listed on the right-hand side. The right axis indicates the voltage V_y in the y direction. Solid circles represent the points used for the scaling analysis in Fig. 5.5. A dashed line labeled $J_x = J_x^*$ shows the power-law behavior. Reprinted from Ref. [1] by S. Maegochi et al. licensed under CC BY 4.0

For these $J_y - E_y$ curves, we determine the depinning current density $J_{y,\mathrm{d}}$ using a 10^{-8} V criterion. The extracted $J_{y,\mathrm{d}}$ is plotted Fig. 5.3a as a function of control parameter J_{x}. The vertical dotted lines are the ordinary depinning current density J_{d} (left) and J_x^* (right). There are three regions; $J_x < J_{\mathrm{d}}$ (region I), $J_{\mathrm{d}} < J_x < J_x^*$ (region II), $J_x > J_x^*$ (region III). In the region I, the vortices under application of J_x are in the pinned state. With superimposing J_y on J_x and increasing J_y, the pinned vortices start to flow obliquely when the vector sum of J_x and J_y exceeds J_{d}. In other words, this is the ordinary depinning in the oblique direction at $J_y = \sqrt{J_{\mathrm{d}}^2 - J_x^2}$ as shown by a quadrant line. In the region II, J_x larger than J_{d} is applied initially, so the vortices flow in the longitudinal direction y before applying J_y. Nevertheless, nonzero value of $J_{y,\mathrm{d}}$ is obtained. This indicates that the nontrivial depinning in the transverse (x) direction occurs, which we call a transverse depinning [18, 19]. In the region III, $J_{y,\mathrm{d}}$ is zero and the vortices can flow freely in the transverse direction by infinitesimal J_y.

5.3.2 Dynamical Ordering Transition

The disappearance of the transverse depinning at $J_x \geq J_x^*$ should be associated with the dynamical transition of the longitudinal vortex flows at J_x^*, where the response to the transverse force changes from a nonlinear behavior with $J_{y,\mathrm{d}} > 0$ (zero linear

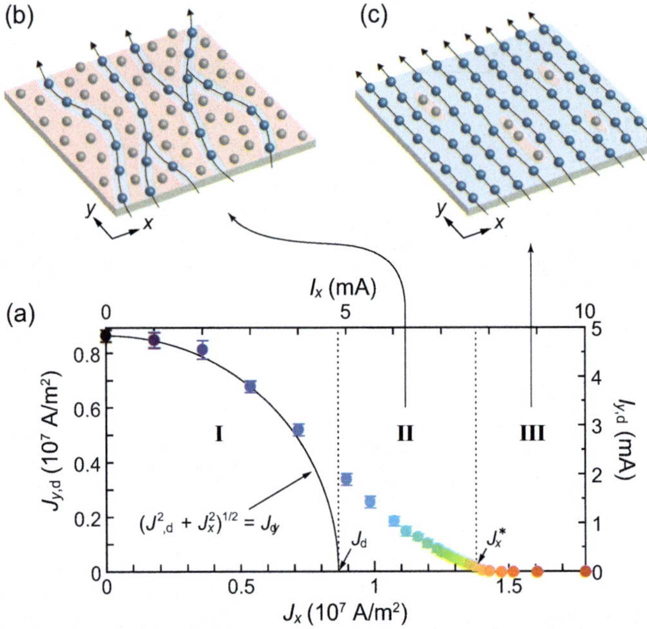

Fig. 5.3 Dynamical phases of vortex flow. **a** $J_{y,\mathrm{d}}$ deduced from the $J_y - E_y$ curves in Fig. 5.2 using a 10^{-8} V criterion plotted against J_x. The right axis indicates the depinning current $I_{y,\mathrm{d}}$ in the y direction. Vertical dotted lines mark the isotropic depinning current density J_d and the threshold current density J_x^*, separating the regions I and II and regions II and III, respectively. In region I, all the vortices are pinned initially and then undergo depinning obliquely, when $(J_y^2 + J_x^2)^{1/2}$ exceeds J_d. A solid quadrant line represents $(J_{y,\mathrm{d}}^2 + J_x^2)^{1/2} = J_\mathrm{d}$. In region II, the vortices initially flowing in the longitudinal (y) direction undergo the transverse depinning for $J_y > J_{y,\mathrm{d}}$. In region III, the transverse depinning does not occur ($J_{y,\mathrm{d}} = 0$). **b**, **c** Schematics of initial vortex flow generated by J_x. **b** The plastic flow in region II. **c** The smectic flow in region III. The areas shaded by red and blue represent the regions with strong and weak effective pinning, respectively. Reprinted from Ref. [1] by S. Maegochi et al. licensed under CC BY 4.0

resistivity) to a linear one with $J_{y,\mathrm{d}} = 0$ (nonzero linear resistivity). In the previous chapter, we have found that the transition or crossover between the plastic and smectic flows occurs around $J_x \approx 1.5 \times 10^7$ A/m^2, which is close to the value of $J_x^* \approx 1.38 \times 10^7$ A/m^2. Therefore, the change of flow states observed at J_x^* is considered to be the dynamical ordering transition from the plastic flow in the region II to the smectic flow in the region III.

Here, we consider a physical picture of the presence and absence of the transverse depinning in the region II and region III, respectively. Figure 5.3b is a schematic illustration of the plastic flow initially generated by $J_x (< J_x^*)$ in the region II. In the plastic flow state, flowing vortices form spatiotemporally fluctuating riverlike structures dominated by pinning, where a small number of vortices flow around islands of pinned vortices shaded by red color [20, 21]. Once such flow structures are formed in the y direction by applying J_x, the flowing vortices shaded by blue color feel transverse barriers in the x direction and the transverse motion by the application of small J_y is less likely. This explains the presence of the transverse depinning in the region II.

In the smectic flow state, in contrast, the effective pinning is sufficiently weakened and the vortices form 1D channels along the longitudinal (y) direction with long-range transverse order in the x direction [22–26] as illustrated in Fig. 5.3c. The reduced effective pinning gives rise to the shrink of the areas of pinned vortices (shaded by red color) and the extension of pinning-free regions (shaded by blue), which percolate in the transverse direction. The vortices consequently feel no transverse barriers and flow freely in the x direction by the application of infinitesimal J_y. This explains the absence of the transverse depinning in the region III.

5.3.3 Scaling Analysis

In this section, we carry out the scaling analysis and reveal the second-order nature of the dynamical ordering transition. The $J_y - E_y$ characteristics in Fig. 5.2b with a sign change of curvature at J_x^* is reminiscent of the isotherms near the critical point for gases [27], Ising ferromagnets [4], and vortex glasses [5–17] and the flow curves for yielding and jamming transitions [28–30]. In these transitions, the second-order nature of the transition has been found through the scaling analysis. Thus, to examine whether the dynamical ordering is the second-order phase transition, we perform the scaling analysis for the $J_y - \rho_y$ curves in Fig. 5.4, which are replotted from the $J_y - E_y$ data with defining the resistivity as $\rho_y \equiv E_y/J_y$.

If the threshold current J_x^* is a critical point of the second-order phase transition, one can assume the following scaling relation [4, 31],

$$\rho_y(J_x, J_y) = |1 - J_x/J_x^*|^{\beta} f_{\pm}\left(\frac{J_y}{|1 - J_x/J_x^*|^{\Delta}}\right). \tag{5.1}$$

Here, β and Δ represent scaling exponents, $w \equiv J_y/|1 - J_x/J_x^*|^{\Delta}$ is the scaling variable, and $f_+(w)$ and $f_-(w)$, respectively, denote branches of the universal scaling function for $J_x > J_x^*$ and $J_x < J_x^*$. For $w \to \infty$, the scaling function asymptotically behaves as $f_{\pm}(w) \approx w^{\beta/\Delta}$. Then, $\rho_y(J_x, J_y)$ takes a power-law form with scale invariance at $J_x = J_x^*$,

$$\rho_y(J_x = J_x^*) \propto J_y^{\beta/\Delta}. \tag{5.2}$$

For $w \to 0$, $f_-(w) = 0$ and $f_+(w) = \text{const.}$, leading to the power-law scaling of a linear resistivity $\rho_{y,\text{lin}}$ for $J_x > J_x^*$

$$\rho_{y,\text{lin}} \propto |1 - J_x/J_x^*|^{\beta}. \tag{5.3}$$

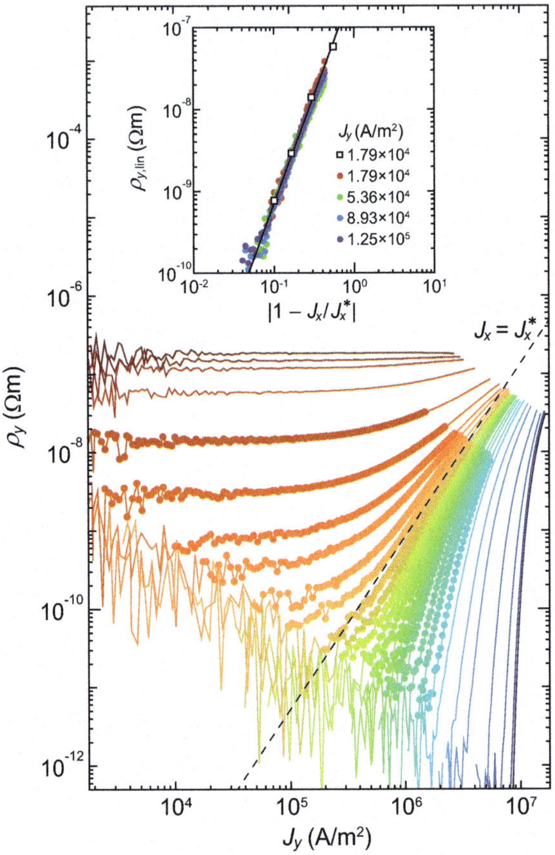

Fig. 5.4 Transverse $J_y - \rho_y$ characteristics. Log-log plots of the $J_y - \rho_y$ data converted from the $J_y - E_y$ data in Fig. 5.2. A dashed line labeled $J_x = J_x^*(\approx 1.38 \times 10^7 \text{ A/m}^2)$ indicates the power law of Eq. 5.2, $\rho_y \propto J_y^{\beta/\Delta}$ with $\beta/\Delta = 2.2 \pm 0.3$. Inset: Log-log plots of the linear resistivity $\rho_{y,\text{lin}}$ as a function of $|1 - J_x/J_x^*|$. Solid circles represent the data from $\rho_y(J_x, J_y)$ measured by using small currents, $J_y = 1.79 \times 10^4$, 5.36×10^4, 8.93×10^4, and 1.25×10^5 A/m², with changing J_x continuously. Open squares are the data collected from $\rho_y(J_x, J_y)$ at $J_y = 1.79 \times 10^4$ A/m² in the main panel. A solid straight line indicates the fit to Eq. 5.3, $\rho_{y,\text{lin}} \propto |1 - J_x/J_x^*|^\beta$ with $\beta = 2.65 \pm 0.3$. Reprinted from Ref. [1] by S. Maegochi et al. licensed under CC BY 4.0

In Fig. 5.4, the power-law scaling of Eq. 5.2 is indeed observed as shown by a dashed line with a slope of $\beta/\Delta = 2.2 \pm 0.3$. In the inset of Fig. 5.4, we plot $\rho_{y,\text{lin}}$ against the dimensionless distance from the critical point, $|1 - J_x/J_x^*|$ in a double logarithmic scale. Open squares are $\rho_{y,\text{lin}}$ extracted from $\rho_y(J_x, J_y)$ at small current density, $J_y = 1.79 \times 10^4$ A/m², in the main panel. Solid circles are $\rho_y(J_x, J_y)$ for small currents, $J_y = 1.79 \times 10^4$, 5.36×10^4, 8.93×10^4, and 1.25×10^5 A/m², measured with continuously changing the control parameter J_x. All data collapse to the straight line that displays the power law in the form of Eq. 5.3 with $\beta = 2.65 \pm 0.3$. Combined with the values of β/Δ and β, we obtain $\Delta = 1.2 \pm 0.2$.

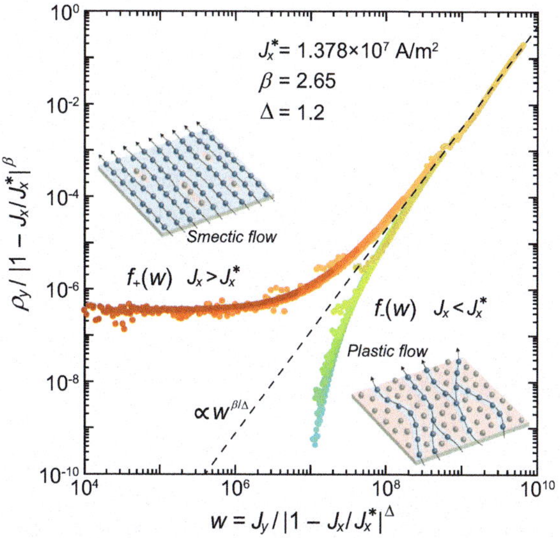

Fig. 5.5 Scaling collapse around the dynamical ordering transition. The scaling plot of the transverse $J_y - \rho_y$ data shown with solid circles in Fig. 5.4 using Eq. 5.1: $\rho_y(J_x, J_y)/|1 - J_x/J_x^*|^\beta$ versus $w(= J_y/|1 - J_x/J_x^*|^\Delta)$. A good collapse of the data to the universal branches is obtained with $J_x^* = 1.378 \times 10^7$ A/m^2, $\beta = 2.65$, and $\Delta = 1.2$. A dashed straight line represents the asymptotic behavior for $w \to \infty$, $\rho_y/|1 - J_x/J_x^*|^\beta \propto w^{\beta/\Delta}$. Inset: Schematics of the plastic flow ($J_x < J_x^*$) and the smectic flow ($J_x > J_x^*$). Reprinted from Ref. [1] by S. Maegochi et al. licensed under CC BY 4.0

Now that we have obtained $J_x^* = 1.378 \times 10^7$ A/m^2, $\beta = 2.65$, and $\Delta = 1.2$, we confirm the critical scaling of Eq. 5.1 for the $J_y - \rho_y$ curves. In Fig. 5.5, we replot the data shown with solid circles in Fig. 5.4 as $\rho_y(J_x, J_y)/|1 - J_x/J_x^*|^\beta$ vs $w(= J_y/|1 - J_x/J_x^*|^\Delta)$. We find a good scaling collapse of the data to universal branches. A dashed straight line represents the asymptotic power-law form $\rho_y/|1 - J_x/J_x^*|^\beta \propto w^{\beta/\Delta}$ for $w \to \infty$. The results reveal that the dynamical ordering transition from the plastic flow to the smectic flow, which takes place at $J_x = J_x^*$, is truly the second-order phase transition. Recent simulation about the Kibble-Zurek mechanism for dynamical ordering also predicts the continuous phase transition [32], which is supported by our experiment in the vortex system show in the next chapter, consistent with the present results.

As mentioned above, the results observed here is analogous to the vortex-glass transition [5–17]. When the control parameter of temperature is lower than the vortex-glass transition temperature, a vortex system settles in the vortex-glass phase dominated by pinning. In this phase, vortices exhibit the depinning transition. For high temperature above the vortex-glass transition, the system is in the vortex-liquid phase, where thermal fluctuations overwhelm the effect of pinning and the depinning is absent. In our work, the disappearance of the transverse depinning at $J_x > J_x^*$ is caused by the reduced effective pinning due to the increased control parameter of J_x.

Given that both parameters act to weaken the effect of pinning, the role of the current J_x in the dynamical ordering transition corresponds to that of the temperature in the vortex-glass transition.

The scaling exponents $\beta = 2.65 \pm 0.3$ and $\Delta = 1.2 \pm 0.2$ obtained here are slightly smaller than those of the vortex-glass transition, $\beta = 4$–8 and $\Delta = 2$–4 [5, 11]. This discrepancy is not surprising because the two transitions are rather different: The dynamical ordering is the nonequilibrium phase transition and the effect of the current is anisotropic while the vortex-glass transition is the equilibrium phase transition and the effect of the temperature is isotropic.

5.4 Conclusion

We explored the response of flowing vortices to the transverse force in the course of the current-induced dynamical ordering by means of the transverse current-voltage measurements. With increasing the control parameter of J_x, the transverse response changed from the nonlinear behavior with the transverse depinning ($J_x < J_x^*$) to the linear one without the transverse depinning ($J_x > J_x^*$), reflecting the dynamical ordering from the disordered plastic flow to the anisotropically ordered smectic flow. We also found the scaling collapse of the transverse $J_y - \rho_y$ curves. The results reveal that the dynamical ordering transition from the plastic flow to the smectic flow is the second-order phase transition.

The transverse force-velocity measurements developed here will be useful to detect the structural change between anisotropic and isotropic structures in other systems such as frictional granular materials, which exhibit the change between a fragile state and a shear-jammed state [33–35].

References

1. Maegochi S, Ienaga K, Okuma S (2024) Evidence of second-order transition and critical scaling for the dynamical ordering transition in current-driven vortices. Sci Rep 14:1232
2. Kadanoff LP, Götze W, Hamblen D, Hecht R, Lewis EAS, Palciauskas VV, Rayl M, Swift J, Aspnes D, Kane J (1967) Static phenomena near critical points: theory and experiment. Rev Mod Phys 39:395
3. Fisher ME (1967) The theory of equilibrium critical phenomena. Rep Prog Phys 30:615
4. Stanley HE (1999) Scaling, universality, and renormalization: Three pillars of modern critical phenomena. Rev Mod Phys 71:S358
5. Koch RH, Foglietti V, Gallagher WJ, Koren G, Gupta A, Fisher MPA (1989) Experimental evidence for vortex-glass superconductivity in Y-Ba-Cu-O. Phys Rev Lett 63:1511
6. Gammel PL, Schneemeyer LF, Bishop DJ (1991) SQUID picovoltometry of $YBa_2Cu_3O_7$ single crystals: evidence for a finite-temperature phase transition in the high-field vortex state. Phys Rev Lett 66:953
7. Olsson HK, Koch RH, Eidelloth W, Robertazzi RP (1991) Observation of critical scaling behavior in the ac impedance at the onset of superconductivity in a large magnetic field. Phys Rev Lett 66:2661

8. Fisher DS, Fisher MPA, Huse DA (1991) Thermal fluctuations, quenched disorder, phase transitions, and transport in type-II superconductors. Phys Rev B 43:130
9. Dekker C, Eidelloth W, Koch RH (1992) Measurement of the exponent μ in the low-temperature phase of $YBa_2Cu_3O_{7-\delta}$ films in a magnetic field: direct evidence for a vortex-glass phase. Phys Rev Lett 68:3347
10. Huse DA, Fisher MPA, Fisher DS (1992) Are superconductors really superconducting? Nature 358:553
11. Roberts JM, Brown B, Hermann BA, Tate J (1994) Scaling of voltage-current characteristics of thin-film Y-Ba-Cu-O at low magnetic fields. Phys Rev B 49:6890
12. Charalambous M, Koch RH, Masselink T, Doany T, Feild C, Holtzberg F (1995) Subpicovolt resolution measurements of the current-voltage characteristics of twinned crystalline $YBa_2Cu_3O_{7-x}$: new evidence for a vortex-glass phase. Phys Rev Lett 75:2578
13. Okuma S, Kokubo N (1997) Universal critical scaling of dc and ac complex resistivities in an indium film near the vortex-glass transition. Phys Rev B 56:14138
14. Okuma S, Arai M (2000) Evidence of the vortex-glass transition in homogeneously disordered thick films of a-Mo_xSi_{1-x}. J Phys Soc Jpn 69:2747
15. Okuma S, Imamoto Y, Morita M (2001) Vortex glass transition and quantum vortex liquid at low temperature in a thick a-Mo_xSi_{1-x} film. Phys Rev Lett 86:3136
16. Okuma S, Morita M, Imamoto Y (2002) Vortex phase diagram and quantum fluctuations in thick a-Mo_xSi_{1-x} films. Phys Rev B 66:104506
17. Okuma S, Togo S, Morita M (2003) Enhancement of the quantum-liquid phase by increased resistivity in thick a-Mo_xSi_{1-x} films. Phys Rev Lett 91:067001
18. Perruchot F, Williams FIB, Mellor CJ, Gaal R, Sas B, Henini M (2000) Transverse threshold for sliding conduction in a magnetically induced Wigner solid. Physica B 284–288:1984
19. Lefebvre J, Hilke M, Altounian Z (2008) Transverse depinning in weakly pinned vortices driven by crossed ac and dc currents. Phys Rev B 78:134506
20. Nori F (1996) Intermittently flowing rivers of magnetic flux. Science 271:1373
21. Matsuda T, Harada K, Kasai H, Kamimura O, Tonomura A (1996) Observation of dynamic interaction of vortices with pinning centers by lorentz microscopy. Science 271:1393
22. Balents L, Marchetti MC, Radzihovsky L (1998) Nonequilibrium steady states of driven periodic media. Phys Rev B 57:7705
23. Le Doussal P, Giamarchi T (1998) Moving glass theory of driven lattices with disorder. Phys Rev B 57:11356
24. Moon K, Scalettar RT, Zimányi GT (1996) Dynamical phases of driven vortex systems. Phys Rev Lett 77:2778
25. Olson CJ, Reichhardt C, Nori F (1998) Nonequilibrium dynamic phase diagram for vortex lattices. Phys Rev Lett 81:3757
26. Kolton AB, Domínguez D, Grønbech-Jensen N (1999) Hall noise and transverse freezing in driven vortex lattices. Phys Rev Lett 83:3061
27. Green MS, Vicentini-Missoni M, Sengers JMHL (1967) Scaling-law equation of state for gases in the critical region. Phys Rev Lett 18:1113
28. Olsson P, Teitel S (2007) Critical scaling of shear viscosity at the jamming transition. Phys Rev Lett 99:178001
29. Yoshino H, Nogawa T, Kim B (2009) Vortex jamming in superconductors and granular rheology. New J Phys 11:013010
30. Paredes J, Michels MAJ, Bonn D (2013) Rheology across the zero-temperature jamming transition. Phys Rev Lett 111:015701
31. Widom B (1965) Equation of state in the neighborhood of the critical point. J Chem Phys 43:3898
32. Reichhardt CJO, del Campo A, Reichhardt C (2022) Kibble-Zurek mechanism for nonequilibrium phase transitions in driven systems with quenched disorder. Commun Phys 5:173
33. Bi D, Zhang J, Chakraborty B, Behringer RP (2011) Jamming by shear. Nature 480:355
34. Zhao Y, Barés J, Zheng H, Socolar JES, Behringer RP (2019) Shear-jammed, fragile, and steady states in homogeneously strained granular materials. Phys Rev Lett 123:158001
35. Behringer RP, Chakraborty B (2019) The physics of jamming for granular materials: a review. Rep Prog Phys 82:012601

Chapter 6
Kibble-Zurek Mechanism for the Dynamical Ordering Transition

6.1 Introduction

We have shown in the previous chapter that the dynamical ordering transition is the second-order phase transition. As explained in Sect. 1.2.3, the Kibble-Zurek mechanism (KZM) is expected to be applicable to continuous second-order phase transitions. Indeed, KZM has been experimentally studied in a wide variety of systems, such as superfluid ^{4}He [1, 2], superfluid ^{3}He [3–7], cosmic microwave background [8, 9], superconductors [10–12], Josephson junctions [13–15], liquid crystals [16, 17], multiferroic materials [18–20], colloidal monolayer [21], nonlinear optical systems [22], convective systems [23, 24], ion coulomb crystals [25–27], ultracold atomic gases [28–31]. Most experiments investigated thermal equilibrium phase transitions, while recent studies have suggested that KZM can be applied to quantum phase transitions [32–36] and nonequilibrium phase transitions [37, 38]. Here, we note that the term "nonequilibrium phase transition" represents the transition between two distinct *nonequilibrium steady states*, where the system is always out of equilibrium. In the context of KZM, the transitions are often referred as nonequilibrium, but it means the equilibrium phase transition between two distinct equilibrium states via *nonequilibrium transient states*, where the phase transition process deviates from the quasi-static condition.

Recent simulation has proposed the applicability of KZM to dynamical ordering transitions between disordered and ordered flow states for many-particle systems driven over disordered substrates [38]. The system referred to in the simulation includes vortices in type-II superconductors. Now, we know that the vortex system exhibits the dynamical ordering transition from the disordered plastic flow to the anisotropically ordered smectic flow with increasing the current. Dynamical ordering is also expected for various systems, including moving Wigner crystals [39], charge-density waves [40], colloidal particles [41], and magnetic skyrmions [42], and thus it is interesting to examine whether KZM can be applied to the dynamical ordering transition in real systems.

S. Maegochi, *Nonequilibrium Phase Transitions in Driven Vortex Matter*, Springer Theses,
https://doi.org/10.1007/978-981-97-2987-6_6

In this chapter, to test the prediction, we experimentally investigate the degree of order of the vortex configuration in the course of the dynamical ordering transition traversed at various quench rate [43]. For this purpose, we employ the two-step measurements consisting of a quench step and a readout step, similar to those used in Chap. 4. We find that the vortex configuration after the quench by ramping I becomes less ordered as the quench rate $\dot{I}$ is increased. In the ordered phase, namely, in the smectic flow state, we estimate the density of topological defects, i.e., dislocations in the vortex lattice, and confirm the power-law scaling of the defect density with the quench rate, which agrees with KZM. We also find the impulse-adiabatic crossover predicted by KZM [44, 45]. Our experimental results verify the applicability of KZM to the nonequilibrium phase transition between disordered and ordered states.

6.2 Experiments

We used the same strip-shaped amorphous Mo_xGe_{1-x} film with weak random pinning as used in Chap. 3. We applied $B = 3.5$ T perpendicular to the film surface at 4.1 K, corresponding to the peak-effect regime. In our film sample, the edge effects on the dynamics of vortices are almost negligible [46].

Figure 6.1 shows the experimental procedure of initialization and following two-step measurements. For initialization, we applied a small current $I = 0.41$ mA to the vortex system for 1 s and prepared the disordered initial state. The value of $I = 0.41$ mA corresponds to the plastic flow regime judged from Fig. 6.2a, where I–V characteristics and corresponding dV/dI are plotted with solid circles and open squares, respectively, and a depinning current is $I_d = 0.29$ mA. Then, in the quench step as the first step, we applied a sawtooth current I to the initial state. Using the sawtooth waveform, the current (driving force) was swept linearly from 0 to the

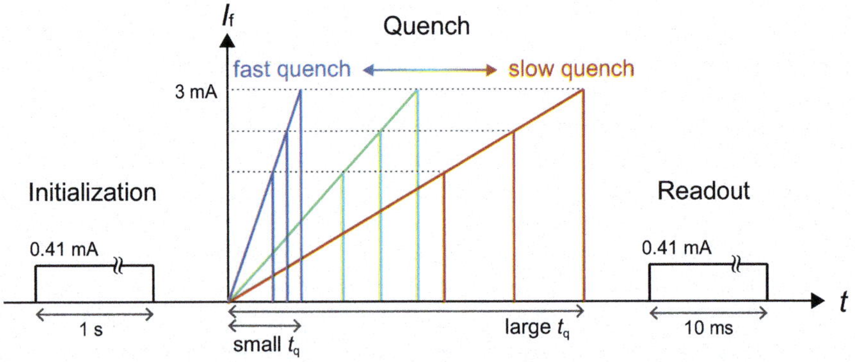

Fig. 6.1 Schematics of the experimental procedure. Initialization and two-step measurements consisting of the quench step and the readout step. Adapted figure with permission from Ref. [43]. Copyright 2022 by the American Physical Society

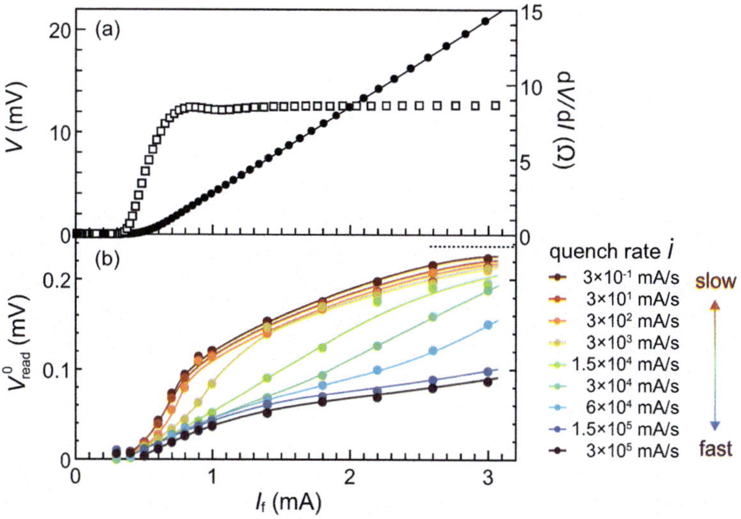

Fig. 6.2 **a** I–V characteristics and the differential resistance dV/dI shown with solid circles and open squares, respectively. **b** The initial value of the voltage obtained in the readout measurement, V^0_{read}. This value reflects the configurational order of vortices at the end of the quench step, where we ramped the current from $I = 0$ to I_{f} with various quench rates $\dot{I}$. The values of $\dot{I}$ are listed on the right-hand side. The dotted line marks the approximate upper bound of $V^0_{\text{read}}(I_{\text{f}}, \dot{I})$, that is, the most ordered configuration in the adiabatic limit, corresponding to $P_6 = 1$. Other lines are drawn as a guide to the eye. Adapted figure with permission from Ref. [43]. Copyright 2022 by the American Physical Society

final value I_{f} over the time of t_{f}. We set the maximum value of I_{f} to 3 mA and specifically defined t_{f} for 3 mA as the quench time t_{q}. We controlled the quench rate, that is, the slope of the sawtooth wave, by changing the ratio $I_{\text{f}}/t_{\text{f}} = 3$ mA/$t_{\text{q}}(\equiv \dot{I})$ for different sets of I_{f} and t_{f}. After shutting off the sawtooth current whose fall time is about 0.7 μs, we performed the second-step readout measurements for evaluating the configurational order of the vortices frozen at the end of the quench. We applied the square current with the amplitude of $I = 0.41$ mA and measured the transient voltage response $V(t)$ using the oscilloscope.

In Chaps. 4 and 5 using the sample similar to that used in this chapter, we found that the dynamical ordering transition between the plastic and smectic flows takes place at a threshold current I_1, which is located around the current where dV/dI takes a half value of its maximum. Therefore, it is considered that the dynamical ordering transition in the sample used here occurs at around $I_1 = 0.56 \pm 0.1$ mA. In our experiments, unlike the thermal equilibrium phase transitions, the dynamics of the vortices and topological defects after the quench can be negligible because at the end of the quench the current is turned off and the vortex system falls into the pinned state.

6.3 Results and Discussion

In Fig. 6.3, we representatively show the readout voltage $V(t)$ after the quench step for $I_f = 3$ mA with various quench rates, $\dot{I} = 3 \times 10^{-1}, 3 \times 10^1, 3 \times 10^2, 3 \times 10^3, 1.5 \times 10^4, 3 \times 10^4, 6 \times 10^4, 1.5 \times 10^5$, and 3×10^5 mA/s from top (dark red) to bottom (dark blue). A larger readout voltage indicates that the vortices are more mobile and their configuration is more ordered. For all data, $V(t)$ decreases monotonically due to a dynamic disordering phenomenon, where vortices are gradually trapped by pinning centers and change from the initially ordered flow state at $t \approx 0$ to the disordered steady state. The initial rise of $V(t)$ at $t \approx 0$ is smaller for larger $\dot{I}$, namely, for the faster quench. This result shows that even when the vortex system was quenched at the same $I_f(> I_1)$, the vortex configuration in the steady state is less ordered for the faster quench, consistent with KZM. Conversely, this consistency indicates that the dynamical ordering transition was indeed traversed during the quench step.

Figure 6.2b shows the initial voltage of the readout measurements, $V^0_{\text{read}}[\equiv V(t = 2\,\mu\text{s})]$, against I_f, where $t = 2\,\mu$s is a response time of the amplifier. The data with the same color are acquired by using the same quench rate $\dot{I}$. The values of $V^0_{\text{read}}(I_f)$ and hence the degree of order of the vortex configuration show a tendency to decrease with an increase in the quench rate, which is in agreement with KZM. For the small quench rates with $\dot{I} = 3 \times 10^{-1}$ and 3×10^1 mA/s, we find that $V^0_{\text{read}}(I_f)$ is almost the same. Therefore, the degree of order of the vortex configuration after the quench remains unchanged when we further decrease the quench rate below $\dot{I} = 3 \times 10^1$ mA/s. Thus, we treat the data for $\dot{I} \leq 3 \times 10^1$ mA/s as the adiabatic limit. The data in this limit exhibits steep rise of V^0_{read} within $I_f = 0.3$–0.9 mA, slow increment above $I_f \approx 1$ mA, and saturation to a value of ≈ 0.24 mV marked by a dotted line.

Now that the vortex configuration has turned out to become less ordered with increasing the quench rate $\dot{I}$, our next step for verifying KZM is to measure the defect density P_D and confirm its scaling with $\dot{I}$. To our knowledge, however, no

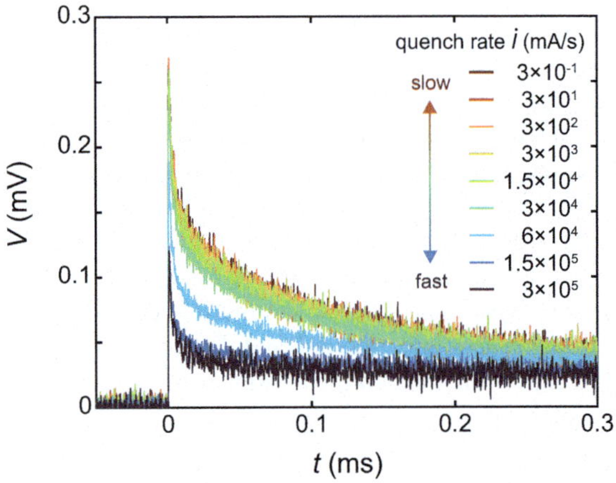

Fig. 6.3 The readout voltage $V(t)$ generated by the motion of vortices subjected to the square current. The readout step was conducted for the vortex configuration after the quench at $I_f = 3$ mA with various $\dot{I}$. The values of $\dot{I}$ are shown in the figure. Adapted figure with permission from Ref. [43]. Copyright 2022 by the American Physical Society

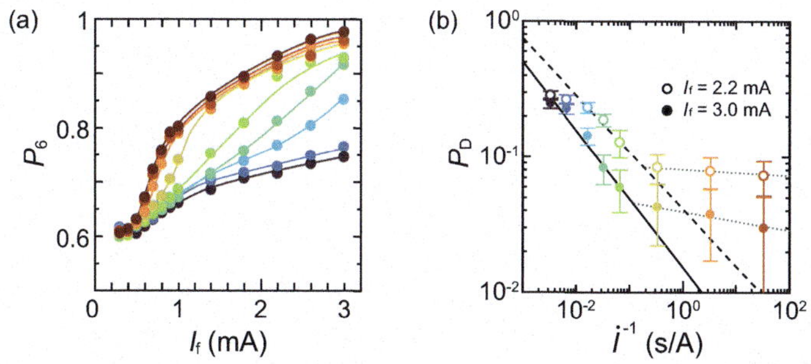

Fig. 6.4 **a** P_6 as a function of I_f deduced from the data shown in Fig. 6.2b. We use the same symbols as those in Fig. 6.2b. The lines are drawn as a guide to the eye. **b** The log-log plot of the defect density $P_D(= 1 - P_6)$ as a function of $\dot{I}^{-1}$ for $I_f = 3$ mA (solid circles) and 2.2 mA (open circles). The solid and dashed lines show the fitting of the data to a power law, $P_D \propto \dot{I}^b$, with $b = 0.5 \pm 0.06$ and $b = 0.42^{+0.06}_{-0.14}$, respectively. The dotted lines are drawn as a guide to the eye. Reprinted figure with permission from Ref. [43]. Copyright 2022 by the American Physical Society

experimental method is available to visualize the configuration of vortices driven by the current in order to measure P_D. Hence, we determine P_D based on the simulation model recently proposed for vortices [38].

In the simulation, the configurational order is evaluated by the fraction of sixfold coordinated vortices P_6 and the defect density is determined from $P_D = 1 - P_6$. The model gives $P_6 \approx 0.6$ just above the depinning force, which indicates the highly defected plastic flow state. As the driving force F_D is increased, P_6 for the adiabatic limit exhibits steep rise and then saturates to a constant value of $P_6 \approx 1$ above the dynamical ordering transition from the disordered plastic flow to the moving anisotropic crystal or the moving smectic flow. With increasing the quench rate, P_6 at a given F_D shows monotonic decrease. Here, we find that $V^0_{\text{read}}(I_f, \dot{I})$ shown in Fig. 6.2b and $P_6(F_D, t_q^{-1})$ in the simulation [38] have very similar dependences on the driving force and the quench rate. Therefore, we regard V^0_{read} akin to P_6 as its approximate representative. We note that in our vortex system, mode-locking experiments demonstrated the presence of a moving triangular lattice of fast driven vortices. The dotted line in Fig. 6.2b marks an approximate maximum of $V^0_{\text{read}}(I_f, \dot{I})$, corresponding to the most ordered configuration with $P_6 = 1$ in the adiabatic limit. Combined with the result of the simulation [38], we determine $P_6 \approx 0.6$ at the depinning I_d, where the configuration of the vortices are most disordered. Then, we deduce the values of $P_6(I_f)$ by means of the linear transformation of V^0_{read} in Fig. 6.2b.

Figure 6.4a shows $P_6(I_f)$ thus obtained for various $\dot{I}$. The $P_6(I_f)$ curves for various $\dot{I}$ shows similar behavior to that of the $P_6(F_D)$ curves for various t_q in the simulation [38]. According to simulation [47], the average vortex velocity (voltage) subjected to a constant ac force shows a linear relation with $P_6(= 1 - P_D)$. We also note that the ambiguity in definition of $P_6 \approx 0.6$ at the depinning I_d and the location of the maximum $V^0_{\text{read}}(I_f, \dot{I})$ does not affect the power-law scaling and the critical exponent in the following discussion.

As explained in Sect. 1.2.3, the defect density P_D is expected to obey Eq. 1.10, namely, $P_\mathrm{D} \propto \dot{I}^b$ in the present case, where $b = (D - d)\nu/(1 + \nu z)$, D and d are the dimensions of the system and the topological defect, respectively, and ν and z are correlation-length and dynamical critical exponents [48] of the underlying phase transition, respectively. To test this KZ scaling, in Fig. 6.4b, we show a log-log plot of P_D for $I_\mathrm{f} = 3.0$ and 2.2 mA with solid and open circles, respectively, as a function of $\dot{I}^{-1}$. For both I_f, P_D follows a decreasing power law with $\dot{I}^{-1}$ and has a tail at larger $\dot{I}^{-1}$, that is, in the adiabatic limit. As shown with the solid and dashed lines, the power law fits are obtained for $I_\mathrm{f} = 3.0$ and 2.2 mA with $b = 0.5 \pm 0.06$ and $b = 0.42^{+0.06}_{-0.14}$, respectively. According to the simulation [38], the dynamical ordering transition from the disordered plastic flow to the ordered smectic flow in the vortex system belongs to the 1(+1) dimensional DP universality class [49], which gives $D = 1$, $d = 0$, $\nu = 1.097$, $z = 1.58$, and $b = 0.401$. Here, the notation "(+1)" is used to explicitly denote the presence of specific direction in the DP model. The values of $b = 0.5 \pm 0.06$ and $b = 0.42^{+0.06}_{-0.14}$ obtained here are close to the value of $b = 0.401$ for the 1(+1) dimensional DP class and $b = 0.39$ found in the simulation [38].

Furthermore, we find another evidence in agreement with KZM from the data shown in Fig. 6.4a. As I_f is increased, $P_6(I_\mathrm{f})$ for larger $\dot{I}$ shows less steep rise around $I_\mathrm{f} \approx 0.5$ mA, deviation from the data in the adiabatic limit ($\dot{I} = 3 \times 10^{-1}$ mA/s) shown with dark red circles, and then tendency to converge on the data in the adiabatic limit at large I_f. Let us focus on the deviation of $P_6(I_\mathrm{f})$ from the ordered configuration in the adiabatic limit for each $\dot{I}$. Figure 6.5a displays the difference of $P_6(I_\mathrm{f})$ for various $\dot{I}$ ($\geq 3 \times 10^1$ mA/s) from $P_6(I_\mathrm{f})[\equiv P_6^{\mathrm{ad}}(I_\mathrm{f})]$ for $\dot{I} = 3 \times 10^{-1}$ mA/s in the adiabatic limit, which we define as $P_6^{\mathrm{ad}}(I_\mathrm{f}) - P_6(I_\mathrm{f})$. Except for the data close to the adiabatic limit, $P_6^{\mathrm{ad}}(I_\mathrm{f}) - P_6(I_\mathrm{f})$ increases steeply against I_f and then takes a peak or a broad maximum at $I_\mathrm{f}(\equiv I_\mathrm{f}^{\mathrm{peak}})$ followed by a decrease toward zero. With an increase in $\dot{I}$, the current $I_\mathrm{f}^{\mathrm{peak}}$ shifts to larger values apart from the threshold current I_1. This trend of $I_\mathrm{f}^{\mathrm{peak}}$ implies the crossover of dynamics from the impulse or frozen-out regime to the adiabatic regime in the ordered phase as predicted by KZM [44, 45].

As seen in Sect. 1.2.3, the impulse-adiabatic crossover point is expressed by Eq. 1.8. In this work, the characteristic current $I_\mathrm{f}^{\mathrm{peak}}$, at which the deviation from the adiabatic limit is maximum because of the freeze-out dynamics, is considered to manifest the impulse-adiabatic crossover. Hence, we expect the following power-law relation,

$$\epsilon(\hat{t}) \propto (I_\mathrm{f}^{\mathrm{peak}} - I_1)/I_1 \propto \dot{I}^{1/(1+\nu z)}, \tag{6.1}$$

where $I_1 (= 0.56 \pm 0.1$ mA) represents a threshold current of the dynamical ordering transition and the reduced control parameter $(I_\mathrm{f}^{\mathrm{peak}} - I_1)/I_1$ corresponds to ϵ in Eq. 1.8. In Fig. 6.5b, we plot $(I_\mathrm{f}^{\mathrm{peak}} - I_1)/I_1$ vs $\dot{I}^{-1}$ in a double logarithmic scale and find the power-law scaling of Eq. 6.1 with $\nu z = 1.7 \pm 0.3$. The product of νz is again in agreement with the theoretical value $\nu z (= \nu_\parallel) \approx 1.73$ for DP universality class

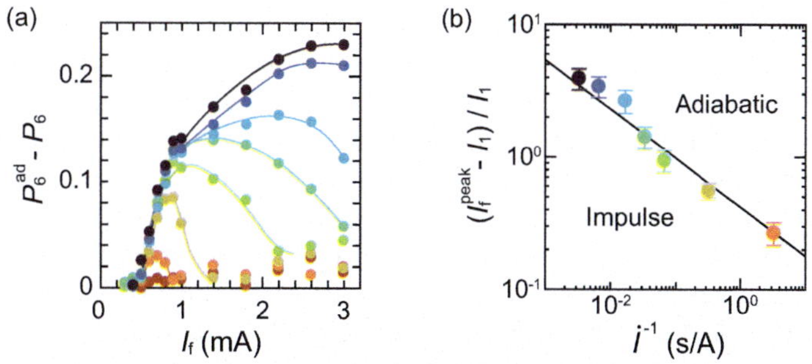

Fig. 6.5 **a** The difference of $P_6(I_f)$ for various $\dot{I}$ ($\geq 3 \times 10^1$ mA/s) from the adiabatic limit $P_6^{ad}(I_f)$. Here, P_6 in the adiabatic limit is determined as P_6 for the smallest quench rate $\dot{I} = 3 \times 10^{-1}$ mA/s and denoted as P_6^{ad}. The lines are drawn as a guide to the eye. **b** The log-log plot of $(I_f^{peak} - I_1)/I_1$ versus $\dot{I}^{-1}$. I_f^{peak} is a current at which $P_6^{ad} - P_6$ in (a) takes the maximum value, and $I_1 (= 0.56\,\text{mA})$ is a threshold current of the dynamical ordering transition from the plastic flow to the smectic flow. The straight line is the power-law fit $(I_f^{peak} - I_1)/I_1 \propto \dot{I}^{1/(1+\nu z)}$ with $\nu z = 1.7 \pm 0.3$. Reprinted figure with permission from Ref. [43]. Copyright 2022 by the American Physical Society

in 1(+1) dimension [49]. The simulation confirmed similar scaling of Eq. 1.7 and obtained $\nu z/(1 + \nu z) = 0.625$ [38]. The value of $\nu z = 1.73$ we find can produce almost identical value, $\nu z/(1 + \nu z) \approx 0.63$.

In this work, we focused on the dynamical ordering transition as a testbed for studying KZM in the nonequilibrium continuous phase transition. Of course, we know other nonequilibrium continuous phase transition, namely, RIT studied in the first half of this thesis. However, we consider that RIT is not suitable for the test of KZM since, to our knowledge, the topological defects are not well-defined in the symmetry-broken (irreversible) phase. We note the fact that the broken symmetry in RIT is a time reversal-symmetry, not a spatial symmetry, and the order parameter (fraction of activity) seems to have no spatial direction. This is stark contrast to the dynamical ordering transition, where spatial symmetry is broken and dislocations are spontaneously formed between uncoupled smectic channels as topological defects. Conversely, if the topological defects can be defined, it should be possible to confirm KZM in other nonequilibrium continuous phase transitions. It is also of interest to investigate the defect formation beyond KZM [50, 51] for nonequilibrium phase transitions.

6.4 Conclusion

We experimentally investigated the configurational order of vortices in the course of the dynamical ordering transition traversed at various quench rates $\dot{I}$. We performed two-step measurements consisting of the quench step and the readout step with transient voltage measurements and found that the vortices were set in less ordered configurations with increasing $\dot{I}$. Combined with the findings of simulation [38], we estimated the defect density P_D, which follows the power-law scaling with $\dot{I}$ predicted by KZM. We also found that as the quench rate $\dot{I}$ is increased, the configurational order of vortices shows deviation from that in the adiabatic limit around the dynamical ordering transition I_1, but tendency to converge deep in the ordered smectic flow state. This result is interpreted as the impulse-adiabatic crossover predicted by KZM [44, 45]. The scaling exponents obtained in the present study are, within errors, consistent with those reported in the simulation [38] and theoretical values for the 1(+1) dimensional DP universality class. Our experimental results demonstrate that KZM can be applied to the nonequilibrium continuous phase transition between disordered and ordered states.

In Chaps. 4–6, we studied the dynamical ordering of dc driven vortices and found that the dynamical ordering from the disordered plastic flow to the ordered smectic flow is the second-order phase transition that belongs to the 1(+1) dimensional DP universality class. In the following, we propose a possible picture of the second-order dynamical ordering transition in terms of the order parameter and correlation length.

First, we consider the order parameter which increases continuously from zero at the dynamical ordering transition I_1 and eventually saturates with increasing the driving force (I). A candidate for the order parameter is the fraction of coupled channels. In the plastic flow regime below the transition I_1, this quantity is of course zero. Above I_1, 1D smectic channels are formed [52–57]. As the driving force increases, these channels gradually couple [55] and the fraction of coupled channels monotonically increases, saturating to unity at I_2, which is the onset of the moving Bragg glass. Another candidate for the order parameter is the fraction of sixfold coordinated flowing vortices, namely, the fraction of the vortex lattice locally formed in the vortex flow. This quantity is similar to the fraction of coupled channels as follows. In the plastic flow regime, the flow fluctuates in time and space, so we cannot identify the stationary formed vortex lattice, and hence the order parameter is zero. On the other hand, in the smectic flow regime, the vortex lattice is locally formed within the coupled channels and its fraction monotonically increases with an increase in the fraction of the coupled channels, saturating to unity in the moving Bragg glass regime. Thus, the fraction of coupled channels and the fraction of the locally formed vortex lattice in motion can be considered as the order parameter that characterizes the smectic flow.

The order parameter probably includes the phase degree of freedom because according to KZM the two-component order parameter is required to consider 0D dislocations in the 2D vortex lattice [58]. A candidate of phase is the lattice orientation. In fact, recent simulation [38] has reported that when the dynamical ordering transition is traversed at the large quench rate, the vortex configuration at the end of the quench is disordered and polycrystalline, where different domains have different lattice orientations. The size of each domain, namely, the phase-coherent region, decreases with an increase in the quench rate, consistent with the KZ scenario. Therefore, the dynamical ordering transition is considered to be characterized by the complex order parameter, whose amplitude is the fraction of coupled channels or the fraction of the locally formed vortex lattice in motion and phase is the lattice orientation.

Next, we consider the correlation length. Since the order parameter is the fraction of coupled channels, the correlation length in the ordered smectic flow phase is considered to be the average lateral size of the region of uncoupled channels or average distance between coupled channels. Upon approaching the critical point I_1 from above, this length scale diverges because all the channels are decoupled at I_1. With an increase in the driving force in the smectic flow phase ($I > I_1$), it becomes smaller because the channels gradually couple, which is consistent with the ordinary description of the correlation length. In the disordered plastic flow phase, the length scale of ordered regions instead of disordered regions gives the correlation length.

Finally, we consider the picture of DP in one dimension. As proposed above, the order parameter is the fraction of coupled channels. If we consider sites of the DP model (e.g., trees in the forest fire model as explained in Sect. 1.1.2) between each adjacent channel, they are aligned in one dimension and the activity (e.g., fire or burning trees in the forest fire model) is defined as the coupling of two adjacent channels. In the smectic flow phase corresponding to the active phase, the location of the coupled channels changes over time, which represents the percolation process of the active sites as shown in Fig. 1.8b–d.

As the future work, it is of interest to study the divergence of the relaxation time at the critical point of the dynamical ordering transition by means of the relaxation measurements used in experiments of RIT [59–63] and the depinning transition [59, 64–66]. For example, it may be useful to perform the critical-quench experiments from the fully ordered (active) moving Bragg glass state.

References

1. Hendry PC, Lawson NS, Lee RAM, McClintock PVE, Williams CDH (1994) Generation of defects in superfluid ^{4}He as an anologue of the formation of cosmic strings. Nature 368:315
2. Dodd ME, Hendry PC, Lawson NS, McClintock PVE, Williams CDH (1998) Nonappearance of vortices in fast mechanical expansions of liquid ^{4}He through the lambda transition. Phys Rev Lett 81:3703
3. Bäuerle C, Bunkov YM, Fisher SN, Godfrin H, Pickett GR (1996) Laboratory simulation of cosmic string formation in the early Universe using superfluid ^{3}He. Nature 382:332
4. Ruutu VMH, Eltsov VB, Gill AJ, Kibble TWB, Krusius M, Makhlin YG, Plaçais B, Volovik GE, Xu W (1996) Vortex formation in neutron-irradiated superfluid ^{3}He as an analogue of cosmological defect formation. Nature 382:334
5. Aranson IS, Kopnin NB, Vinokur VM (1999) Nucleation of vortices by rapid thermal quench. Phys Rev Lett 83:2600
6. Aranson IS, Kopnin NB, Vinokur VM (2001) Dynamics of vortex nucleation by rapid thermal quench. Phys Rev B 63:184501
7. Eltsov VB, Hänninen R, Krusius M (2010) Vortices and other topological defects in non-equilibrium phase transitions of superfluid ^{3}He. Physica C 470:803
8. Coulson D, Ferreira P, Graham P, Turok N (1994) Microwave anisotropies from cosmic defects. Nature 368:27
9. Bevis N, Hindmarsh M, Kunz M, Urrestilla J (2008) Fitting cosmic microwave background data with cosmic strings and inflation. Phys Rev Lett 100:021301

10. Maniv A, Polturak E, Koren G (2003) Observation of magnetic flux generated spontaneously during a rapid quench of superconducting films. Phys Rev Lett 91:197001
11. Polturak E (2019) Beyond the horizon: magneto-optical imaging studies of the kibble-zurek scenario in superconductors. J Low Temp Phys 197:310
12. Rochet A, Vadimov V, Magrini W, Thakur S, Trebbia J-B, Melnikov A, Buzdin A, Tamarat P, Lounis B (2020) On-demand optical generation of single flux quanta. Nano Lett 20:6488
13. Carmi R, Polturak E, Koren G (2000) Observation of spontaneous flux generation in a multi-josephson-junction loop. Phys Rev Lett 84:4966
14. Monaco R, Mygind J, Rivers RJ (2002) Zurek-kibble domain structures: the dynamics of spontaneous vortex formation in annular josephson tunnel junctions. Phys Rev Lett 89:080603
15. Monaco R, Mygind J, Aaroe M, Rivers RJ, Koshelets VP (2006) Zurek-kibble mechanism for the spontaneous vortex formation in Nb-Al/Al_{ox}/Nb josephson tunnel junctions: new theory and experiment. Phys Rev Lett 96:180604
16. Chuang I, Durrer R, Turok N, Yurke B (1991) Cosmology in the laboratory: defect dynamics in liquid crystals. Science 251:1336
17. Bowick MJ, Chandar L, Schiff EA, Srivastava AM (1994) The cosmological kibble mechanism in the laboratory: string formation in liquid crystals. Science 263:943
18. Griffin SM, Lilienblum M, Delaney KT, Kumagai Y, Fiebig M, Spaldin NA (2012) Scaling behavior and beyond equilibrium in the hexagonal manganites. Phys Rev X 2:041022
19. Lin S-Z, Wang X, Kamiya Y, Chern G-W, Fan F, Fan D, Casas B, Liu Y, Kiryukhin V, Zurek WH, Batista CD, Cheong S-W (2014) Topological defects as relics of emergent continuous symmetry and Higgs condensation of disorder in ferroelectrics. Nat Phys 10:970
20. Meier QN, Lilienblum M, Griffin SM, Conder K, Pomjakushina E, Yan Z, Bourret E, Meier D, Lichtenberg F, Salje EKH, Spaldin NA, Fiebig M, Cano A (2017) Global formation of topological defects in the multiferroic hexagonal manganites. Phys Rev X 7:041014
21. Deutschländer S, Dillmann P, Maret G, Keim P (2015) Kibble-Zurek mechanism in colloidal monolayers. Proc Natl Acad Sci USA 112:6925
22. Ducci S, Ramazza PL, González-Viñas W, Arecchi FT (1999) Order parameter fragmentation after a symmetry-breaking transition. Phys Rev Lett 83:5210
23. Casado S, González-Viñas W, Mancini H (2006) Testing the Kibble-Zurek mechanism in Rayleigh-Bénard convection. Phys Rev E 74:047101
24. Casado S, González-Viñas W, Boccaletti S, Ramazza PL, Mancini H (2007) The birth of defects in pattern formation: Testing of the Kibble-Zurek mechanism. Eur Phys J Special Topics 146:87
25. del Campo A, De Chiara G, Morigi G, Plenio MB, Retzker A (2010) Structural defects in ion chains by quenching the external potential: the inhomogeneous kibble-zurek mechanism. Phys Rev Lett 105:075701
26. Ulm S, RoBnagel J, Jacob G, Degünther C, Dawkins ST, Poschinger UG, Nigmatullin R, Retzker A, Plenio MB, Schmidt-Kaler F, Singer K (2013) Observation of the Kibble-Zurek scaling law for defect formation in ion crystals. Nat Commun 4:2290
27. Pyka K, Keller J, Partner HL, Nigmatullin R, Burgermeister T, Meier DM, Kuhlmann K, Retzker A, Plenio MB, Zurek WH, del Campo A, Mehlstäubler TE (2013) Topological defect formation and spontaneous symmetry breaking in ion Coulomb crystals. Nat Commun 4:2291
28. Sadler LE, Higbie JM, Leslie SR, Vengalattore M, Stamper-Kurn DM (2006) Spontaneous symmetry breaking in a quenched ferromagnetic spinor Bose-Einstein condensate. Nature 443:312
29. Weiler CN, Neely TW, Scherer DR, Bradley AS, Davis MJ, Anderson BP (2008) Spontaneous vortices in the formation of Bose-Einstein condensates. Nature 455:948
30. Navon N, Gaunt AL, Smith RP, Hadzibabic Z (2015) Critical dynamics of spontaneous symmetry breaking in a homogeneous Bose gas. Science 347:167
31. Ko B, Park JW, Shin Y (2019) Kibble-Zurek universality in a strongly interacting Fermi superfluid. Nat Phys 15:1227
32. Dziarmaga J (2010) Dynamics of a quantum phase transition and relaxation to a steady state. Adv Phys 59:1063

33. Chen D, White M, Borries C, DeMarco B (2011) Quantum quench of an atomic mott insulator. Phys Rev Lett 106:235304
34. Kolodrubetz M, Clark BK, Huse DA (2012) Nonequilibrium dynamic critical scaling of the quantum ising chain. Phys Rev Lett 109:015701
35. Clark LW, Feng L, Chin C (2016) Universal space-time scaling symmetry in the dynamics of bosons across a quantum phase transition. Science 354:606
36. Keesling A, Omran A, Levine H, Bernien H, Pichler H, Choi S, Samajdar R, Schwartz S, Silvi P, Sachdev S, Zoller P, Endres M, Greiner M, Vuletić V, Lukin MD (2019) Quantum Kibble-Zurek mechanism and critical dynamics on a programmable Rydberg simulator. Nature 568:207
37. Zamora A, Dagvadorj G, Comaron P, Carusotto I, Proukakis NP, Szymańska MH (2020) Kibble-Zurek mechanism in driven dissipative systems crossing a nonequilibrium phase transition. Phys Rev Lett 125:095301
38. Reichhardt CJO, del Campo A, Reichhardt C (2022) Kibble-Zurek mechanism for nonequilibrium phase transitions in driven systems with quenched disorder. Commun Phys 5:173
39. Reichhardt C, Olson CJ, Grønbech-Jensen N, Nori F (2001) Moving wigner glasses and smectics: dynamics of disordered wigner crystals. Phys Rev Lett 86:4354
40. Danneau R, Ayari A, Rideau D, Requardt H, Lorenzo JE, Ortega L, Monceau P, Currat R, Grübel G (2002) Motional ordering of a charge-density wave in the sliding state. Phys Rev Lett 89:106404
41. Tierno P (2012) Depinning and collective dynamics of magnetically driven colloidal monolayers. Phys Rev Lett 109:198304
42. Reichhardt C, Ray D, Reichhardt CJO (2015) Collective transport properties of driven skyrmions with random disorder. Phys Rev Lett 114:217202
43. Maegochi S, Ienaga K, Okuma S (2022) Kibble-zurek mechanism for dynamical ordering in a driven vortex system. Phys Rev Lett 129:227001
44. Zurek WH (1985) Cosmological experiments in superfluid helium? Nature 317:505
45. del Campo A, Kibble TWB, Zurek WH (2013) Causality and non-equilibrium second-order phase transitions in inhomogeneous systems. J Phys Condens Matter 25:404210
46. Okuma S, Kashiro K, Suzuki Y, Kokubo N (2008) Order-disorder transition of vortex matter in a-Mo_xGe_{1-x} films probed by noise. Phys Rev B 77:212505
47. Valenzuela SO (2002) Order and mobility of solid vortex matter in oscillatory driving currents. Phys Rev Lett 88:247003
48. Hohenberg PC, Halperin BI (1977) Theory of dynamic critical phenomena. Rev Mod Phys 49:435
49. Hinrichsen H (2000) Non-equilibrium critical phenomena and phase transitions into absorbing states. Adv Phys 49:815
50. Goo J, Lim Y, Shin Y (2021) Defect saturation in a rapidly quenched bose gas. Phys Rev Lett 127:115701
51. Goo J, Lee Y, Lim Y, Bae D, Rabga T, Shin Y (2022) Universal early coarsening of quenched bose gases. Phys Rev Lett 128:135701
52. Balents L, Marchetti MC, Radzihovsky L (1998) Nonequilibrium steady states of driven periodic media. Phys Rev B 57:7705
53. Le Doussal P, Giamarchi T (1998) Moving glass theory of driven lattices with disorder. Phys Rev B 57:11356
54. Moon K, Scalettar RT, Zimányi GT (1996) Dynamical phases of driven vortex systems. Phys Rev Lett 77:2778
55. Olson CJ, Reichhardt C, Nori F (1998) Nonequilibrium dynamic phase diagram for vortex lattices. Phys Rev Lett 81:3757
56. Kolton AB, Domínguez D, Grønbech-Jensen N (1999) Hall noise and transverse freezing in driven vortex lattices. Phys Rev Lett 83:3061
57. Fangohr H, Cox DJ, de Groot PAJ (2001) Vortex dynamics in two-dimensional systems at high driving forces. Phys Rev B 64:064505
58. Kibble TWB (1976) Topology of cosmic domains and strings. J Phys A Math Gen 9:1387

59. Okuma S, Tsugawa Y, Motohashi A (2011) Transition from reversible to irreversible flow: Absorbing and depinning transitions in a sheared-vortex system. Phys Rev B 83:012503
60. Dobroka M, Kawamura Y, Ienaga K, Kaneko S, Okuma S (2017) Memory formation and evolution of the vortex configuration associated with random organization. New J Phys 19:053023
61. Maegochi S, Ienaga K, Kaneko S, Okuma S (2019) Critical behavior near the reversible-irreversible transition in periodically driven vortices under random local shear. Sci Rep 9:16447
62. Miyagawa K, Maegochi S, Ienaga K, Kaneko S, Okuma S (2021) Effects of the velocity on the reversible-irreversible transition in a periodically sheared vortex system. J Phys Conf Ser 1975:012002
63. Maegochi S, Ienaga K, Okuma S (2021) Critical behavior of density-driven and shear-driven reversible-irreversible transitions in cyclically sheared vortices. Sci Rep 11:19280
64. Okuma S, Motohashi A (2012) Critical behavior associated with transient dynamics near the depinning transition. New J Phys 14:123021
65. Kawamura Y, Moriya S, Ienaga K, Kaneko S, Okuma S (2017) Nonequilibrium depinning transition of ac driven vortices with random pinning. New J Phys 19:093001
66. Kaji T, Maegochi S, Ienaga K, Kaneko S, Okuma S (2022) Critical behavior of nonequilibrium depinning transitions for vortices driven by current and vortex density. Sci Rep 12:1542

Chapter 7
Conclusions

In this thesis, we have experimentally investigated the nonequilibrium phase transitions in driven vortex matter. In recent years, there has been a surge of interest in nonequilibrium phase transitions, while they are still in infancy compared to the study on equilibrium phase transitions. To further understand them, more exploration, especially the experimental study is needed. We study the nonequilibrium phase transitions in externally driven many-particle systems by ac and dc forces, that is, the reversible-irreversible transition (RIT) and the dynamical ordering transition, respectively. We used the vortex system of the amorphous Mo_xGe_{1-x} film, which is a conventional superconductor with random point pinning, thus free from complicated effects that might affect transport properties and vortex dynamics, such as special band structures, specific crystal symmetry, extended pinning, and edge contamination effects. Therefore, our findings are of fundamental and general importance, and universal.

RIT was discovered in colloidal experiments in 2008 and then its universality was proved by our group using vortex systems. In these experiments global periodic shear was used. However, recently, we found that RIT was caused by random local shear instead of the artificial global shear. In Chap. 3, we studied RIT driven by particle (vortex) density as well as shear amplitude [1]. We performed shearing experiments by changing the magnetic field B with fixed shear amplitude d and observed the power-law divergence of the relaxation time $\tau(B)$ at RIT. The critical exponents $\alpha = 0.45 \pm 0.05$ and $\nu = 1.32 \pm 0.10$ for the density-driven RIT are, within errors, in agreement with $\alpha = 0.40 \pm 0.05$ and $\nu = 1.38 \pm 0.08$, respectively, for the shear-driven RIT [2]. The results show that RIT is universal against the control parameters B and d and the universality classes of both types of RIT are the same as the directed percolation (DP) in two dimensions (2D), which predicts $\alpha \approx 0.45$ and $\nu = 1.295 \pm 0.006$ [3]. We also found the transition or the crossover between two distinct reversible states in the shear-driven experiment, corresponding to the collisionless point-reversible state $d < d_{c1} (\approx a_0)$, where a_0 is an intervortex spacing, and the loop-reversible state ($d_{c1} < d < d_c$) with vortex-vortex collisions. In either

S. Maegochi, *Nonequilibrium Phase Transitions in Driven Vortex Matter*, Springer Theses,
https://doi.org/10.1007/978-981-97-2987-6_7

reversible regime, $\tau(d)$ exhibits a power-law divergence at the same d_c with the almost identical exponents: $\nu = 1.38$ for the loop-reversible state and $\nu = 1.30$ for the point-reversible state. To our knowledge, this is the first observation of the point-reversible and loop-reversible states in the same system. We expect that this work will stimulate further experiments on RIT in different systems. In dilute systems such as colloidal suspensions where the point-reversible state has been observed, the loop-reversible state will be found in the vicinity of RIT, while in dense systems such as soft glasses where the loop-reversible state has been observed, the point-reversible state will be found in the reversible state far away from RIT.

In Chaps. 4–6, we studied the dynamical ordering of dc driven vortices. The dynamical ordering of many interacting particles driven over a disordered substrate has been extensively studied in a wide variety of systems, particularly in the vortex system. The successive dynamical orderings from the plastic flow to the anisotropic smectic flow, and from the smectic flow to the isotropic moving Bragg glass have been numerically studied. However, they have not been verified by any experiment.

In Chap. 4, we investigated the dynamical ordering as a function of dc current I [4]. We prepared the cross-shaped sample and developed the two-step measurements consisting of mutually perpendicular drives (I). Using this method, we examined the transverse response of vortices, which were first driven in one (longitudinal) direction and then frozen in the steady state. We found that the vortices subjected to the second drive in the transverse direction are more mobile under small driving force, less mobile under intermediate driving force, and almost equally mobile under sufficiently large driving force, compared to the steady-state flow in the first drive. This indicates the dynamical ordering from the plastic flow with directional fragility due to the clogging to the anisotropic smectic flow with directional robustness due to the transverse order, and that from the anisotropic smectic flow to the isotropic moving Bragg glass. We also developed the transverse mode-locking (ML) measurements. The observed larger transverse ML resonance than the longitudinal ML provides convincing evidence of the moving smectic phase with higher transverse order than the longitudinal one. We consider that the transverse ML developed here combined with the longitudinal ML is a powerful and convenient tool to detect the anisotropic periodicity in fast driven media, where visualization techniques are unavailable, such as colloids, Wigner solids, and skyrmions.

In Chap. 5, we developed measurements of transverse current-voltage (force-velocity) characteristics and demonstrated that the dynamical ordering transition from the plastic flow to anisotropically ordered smectic flow is the second-order phase transition [5]. We measured the response of longitudinally moving vortices, instead of frozen vortices studied in the previous chapter, to the transverse drive. From transverse current-voltage characteristics under various longitudinal currents superimposed on the transverse current, we found a continuous change of the transverse response from a nonlinear behavior with nonzero transverse depinning current (zero linear resistivity) to a linear behavior with zero transverse depinning current (nonzero linear resistivity) at a well-defined longitudinal current, which corresponds to the dynamical ordering from the plastic to smectic flows. Furthermore, the scaling analysis successfully showed that the transverse current-voltage curves collapse to a universal scaling function, thus providing direct evidence of the second-order dynamical ordering transition.

Finally, in Chap. 6, we tested the applicability of the Kibble-Zurek mechanism (KZM) to the dynamical ordering transition [6]. Using two-step measurements consisting of the quench step and the readout step, we examined the configurational order of vortices in the course of the dynamical ordering for various sweep rates of I, namely, various quench rates $\dot{I}$. We found that the final vortex configuration subjected to faster quench is less ordered. Combined with the results of simulation, we estimated the defect density P_{D} and found the power-law scaling of P_{D} with $\dot{I}$, consistent with the KZ prediction. We further found the crossover from the impulse regime around the critical point to the adiabatic regime where the time evolution of the system resumes, which is also the key prediction by KZM. The scaling exponents obtained in this work are, within errors, consistent with those reported in the recent simulation and theoretical values for the $1(+1)$ dimensional DP universality class. While KZM has been extensively studied for equilibrium phase transitions since 1970's, our results provide first experimental evidence for the applicability of KZM to the nonequilibrium phase transition between disordered and ordered flow states. We expect that KZM will be confirmed in other nonequilibrium continuous phase transitions.

Through this thesis, we revealed that both RIT and the dynamical ordering transition belong to the same universality class as DP, which is the fundamental universality class of nonequilibrium phase transitions. Considering that experimental examples of DP have been reported independently in limited systems, as mentioned in Sect. 1.1.2, the observation of DP for RIT and the dynamical ordering transition in the same system is an important contribution to the study of nonequilibrium phase transitions. In the Tables 7.1 and 7.2, the critical exponents obtained in this thesis and those reported theoretically for the DP class are summarized. The findings obtained in this thesis are not limited to the vortex system but generalized to various many-particle systems. We expect that the findings and experimental methods presented in this thesis stimulate further research on nonequilibrium phase transitions and the nonequilibrium Kibble-Zurek mechanism in a broad class of systems, including colloids, emulsions, soft glasses, dislocations, foams, jammed materials, amorphous solids, spin glasses, charge-density waves, Wigner solids, magnetic domain walls, skyrmions, and active matter.

Table 7.1 Critical exponents for RIT obtained in Chap. 3, which are compared with theoretical ones for 2D DP [3]

Critical exponent	$D = 2$	Shear-driven RIT [2]	Density-driven RIT [1]
$\nu_{\parallel}$	1.295(6)	1.38 ± 0.08	1.32 ± 0.10
$\delta = \alpha$	0.451	0.40 ± 0.05	0.45 ± 0.05

Table 7.2 Critical exponents for the dynamical ordering transition obtained in Chap. 6, which are compared with theoretical ones for 1D DP [3]

Critical exponent	$D = 1$	Dynamical ordering transition [6]
$\nu_\perp (= \nu)$	1.096854(4)	
$\nu_\parallel (= \nu z)$	1.733847(6)	1.7 ± 0.3
z	1.580745(10)	
$b = \frac{(D-d)\nu}{1+\nu z}$	0.401	0.4–0.5

References

1. Maegochi S, Ienaga K, Okuma S (2021) Critical behavior of density-driven and shear-driven reversible-irreversible transitions in cyclically sheared vortices. Sci Rep 11:19280
2. Maegochi S, Ienaga K, Kaneko S, Okuma S (2019) Critical behavior near the reversible-irreversible transition in periodically driven vortices under random local shear. Sci Rep 9:16447
3. Hinrichsen H (2000) Non-equilibrium critical phenomena and phase transitions into absorbing states. Adv Phys 49:815
4. Maegochi S, Ienaga K, Okuma S (2022) Moving smectic phase and transverse mode locking in driven vortex matter. Phys Rev Res 4:033085
5. Maegochi S, Ienaga K, Okuma S (2024) Evidence of second-order transition and critical scaling for the dynamical ordering transition in current-driven vortices. Sci Rep 14:1232
6. Maegochi S, Ienaga K, Okuma S (2022) Kibble-zurek mechanism for dynamical ordering in a driven vortex system. Phys Rev Lett 129:227001

Curriculum Vitae

Shun Maegochi

Fusion and Accelerators Department, Fusion and Superconducting Equipment Design Group Hitachi Ltd.
3-1-1, Saiwai-cho, Hitachi, 317-8511 Japan

Appointments

- Specialist, Hitachi Ltd. (April 2023–present)
- JSPS research fellow (DC1), Department of Physics, Tokyo Institute of Technology (April 2020–March 2023)

Education

- Ph.D. in Science, Tokyo Institute of Technology (April 2020–March 2023) Supervisor: Prof. Satoshi Okuma
- MSc, Tokyo Institute of Technology (April 2018–March 2020)
- BSc, Tokyo Institute of Technology (April 2014–March 2018)

Peer-Reviewed Papers

1. S. Maegochi, K. Ienaga, and S. Okuma
   "Evidence of second-order transition and critical scaling for the dynamical ordering transition in current-driven vortices"
   Sci. Rep. **14**, 1232 (2024).

S. Maegochi, *Nonequilibrium Phase Transitions in Driven Vortex Matter*, Springer Theses,
https://doi.org/10.1007/978-981-97-2987-6

2. S. Maegochi, K. Ienaga, and S. Okuma
   "Kibble-Zurek Mechanism for Dynamical Ordering in a Driven Vortex System"
   Phys. Rev. Lett. **129**, 227001 (2022).
3. S. Maegochi, K. Ienaga, and S. Okuma
   "Moving smectic phase and transverse mode locking in driven vortex matter"
   Phys. Rev. Research **4**, 033085 (2022).
4. T. Kaji, S. Maegochi, K. Ienaga, S. Kaneko, and S. Okuma
   "Critical behavior of nonequilibrium depinning transitions for vortices driven by current and vortex density"
   Sci. Rep. **12**, 1542 (2022).
5. R. Ueno, K. Ienaga, S. Maegochi, H. Shishido, and S. Okuma
   "Dc vortex-flow resistivity in an epitaxially grown MgB2 film probed by pulsed current measurements"
   J. Phys.: Conf. Ser. **2323**, 012013 (2022).
6. S. Maegochi, K. Ienaga, and S. Okuma
   "Critical behavior of density-driven and shear-driven reversible–irreversible transitions in cyclically sheared vortices"
   Sci. Rep. **11**, 19280 (2021).
7. K. Miyagawa, S. Maegochi, K. Ienaga, S. Kaneko, and S. Okuma
   "Effects of the velocity on the reversible-irreversible transition in a periodically sheared vortex system"
   J. Phys.: Conf. Ser. **1975**, 012002 (2021).
8. S. Maegochi, K. Ienaga, S. Kaneko, and S. Okuma
   "Critical behavior near the reversible-irreversible transition in periodically driven vortices under random local shear"
   Sci. Rep. **9**, 16447 (2019).
9. S. Maegochi, M. Dobroka, K. Ienaga, S. Kaneko, and S. Okuma
   "Time evolution of the vortex configuration associated with dynamic ordering detected by dc drive"
   J. Phys.: Conf. Ser. **1293**, 012023 (2019).

Awards

1. Seiichi Tejima Doctoral Dissertation Award
   Shun Maegochi
   "Nonequilibrium phase transitions in driven vortex matter: the reversible-irreversible transition, dynamical ordering, and Kibble-Zurek mechanism," Mar. 2024.
2. Young Scientist Award of the Physical Society of Japan (Division 6)
   Shun Maegochi
   "Experimental study of nonequilibrium phase transitions using a superconducting vortex system," Nov. 2023.

3. Springer Theses Award
   Shun Maegochi
   "Nonequilibrium Phase Transitions in Driven Vortex Matter – The Reversible-Irreversible Transition, Dynamical Ordering, and Kibble-Zurek Mechanism," Mar. 2023.
4. Editors' suggestion (Physical Review Letters)
   Shun Maegochi, Koichiro Ienaga, and Satoshi Okuma
   "Kibble-Zurek Mechanism for Dynamical Ordering in a Driven Vortex System," Nov. 2022.
5. Student Presentation Award of the Physical Society of Japan
   Shun Maegochi
   "Scaling analysis for the transverse depinning transition," Sep. 2022.
6. Student Presentation Award of the Physical Society of Japan
   Shun Maegochi
   "Transverse depinning transition in a superconducting vortex system," Mar. 2022.
7. ISS2021 Encouragement Award
   Shun Maegochi
   "Observation of two reversible regimes in cyclically sheared vortices," Dec. 2021.
8. Student Presentation Award of the Physical Society of Japan
   Shun Maegochi
   "Transverse mode-locking resonance in a superconducting vortex system," Sep. 2021.
9. Student Presentation Award of the Physical Society of Japan
   Shun Maegochi
   "Directional memory effect in the cross driven vortex system," Mar. 2021.
10. 2019 Outstanding Master Student Award in Department of Physics (Tokyo Institute of Technology)
    Shun Maegochi
    "Reversible-irreversible transition in periodically sheared vortices under random local shear," Feb. 2020.
11. Vortex Forum Young Scientist Presentation Award
    Shun Maegochi
    "Critical behavior of reversible-irreversible transitions in a superconducting vortex system," Dec. 2019.
12. Student Presentation Award of the Physical Society of Japan
    Shun Maegochi
    "Influences of vortex density on the reversible-irreversible transition," Sep. 2019.
13. Student Presentation Award of the Physical Society of Japan
    Shun Maegochi
    "Phase diagram of reversible-irreversible transition in a superconducting vortex system," Mar. 2019.

SPRINGER NATURE

GPSR Compliance

The European Union's (EU) General Product Safety Regulation (GPSR) is a set of rules that requires consumer products to be safe and our obligations to ensure this.

If you have any concerns about our products, you can contact us on ProductSafety@springernature.com

In case Publisher is established outside the EU, the EU authorized representative is:

Springer Nature Customer Service Center GmbH
Europaplatz 3
69115 Heidelberg, Germany

The manufacturer's authorised representative in the EU is Springer
Nature Customer Service Centre GmbH, Europaplatz 3, 69115 Heidelberg,
Germany. If you have any concerns regarding our products, please
contact ProductSafety@springernature.com

Printed and bound by CPI Group (UK) Ltd, Croydon, CR0 4YY
01/10/2026
02225877-0001